INFORMATION SYSTEMS ANALYSIS
with an Introduction to Fourth-Generation Technologies

V.J. Hall
University of Waterloo

J.W. Mosevich
Merrill Lynch

Prentice-Hall Canada Inc., Scarborough, Ontario

Canadian Cataloguing in Publication Data

Hall, Vicki J., 1947-
Information systems analysis

Includes bibliographical references.
ISBN 0-13-464363-1

1. Management information systems. 2. System analysis. 3. System design. I. Mosevich, Jack Walter, 1944- II. Title.

T 58.6.H34 1987 658.4'032 C87-094796-6

Prentice-Hall Inc., Englewood Cliffs, *New Jersey*
Prentice-Hall International, Inc., *London*
Prentice-Hall of Australia, Pty., Ltd., *Sydney*
Prentice Hall of India Pvt., Ltd., *New Delhi*
Prentice-Hall of Japan, Inc., *Tokyo*
Prentice-Hall of Southeast Asia (PTE.) Ltd., *Singapore*
Editora Prentice-Hall do Brasil Ltda., *Rio de Janeiro*
Prentice-Hall Hispanoamerica, S.A., *Mexico*

ISBN 0-13-464363-1

Cover photograph: Reproduced with the kind permission of Honeywell Bull Ltd.
Coordinating editor: Edward O'Connor
Production editor: Susan Howlett
Manufacturing buyer: Matt Lumsdon

Printed and bound in Canada by The Alger Press Ltd.
1 2 3 4 5 AP 92 91 90 89 88

TABLE OF CONTENTS

PREFACE

Systems analysis concerns the process of creating a conceptual design of computer software. As currently practised it is both an art and a science. With this in mind, the goal of this text is to teach the knowledge and skills needed for the understanding and practice of systems analysis. The software industry is rapidly evolving, and this evolution is closely coupled with a rise in the importance of information systems. We have chosen to present the most popular traditional approach (structured analysis using data flow diagrams), followed by coverage of prototyping and fourth-generation technologies. This approach has led to a natural organization of the text into three parts.

Part 1 establishes the business context for information systems. Part 2 covers traditional approaches, which are still utilized in the vast majority of companies, and Part 3 introduces fourth-generation technologies, prototyping and end-user computing. There is a tendency today to expect artificial intelligence and fifth-generation computers to automate software development. Although this is being worked on, there are few, if any, systems available, and it is not clear when there will be. Software development is still a manual, labor-intensive process.

In addition, because software is developed for a specific business or functional purpose, there is always a need to understand the particular underlying processes that the software is supposed to improve. This aspect of analysis is a major theme throughout the text. Much of the insight as to how analysis is performed, including psychological and political problems, is based on the authors' experiences in major companies.

It is important to note that we devote roughly one third of the text to the traditional approach, despite the fact that it has many faults and limitations. This is because i) it will be utilized for a long time to come, ii) there is no proven alternative yet available, and iii) because many of the tools used in the traditional approach are appropriate for development with fourth-generation technologies.

ACKNOWLEDGMENTS

We would like to thank the following reviewers for their comments on early drafts of the manuscript: F. H. Lochovsky, University of Toronto: David Haisell, Humber College; Ron Archibald, Southern Alberta Institute of Technology; B. L. Ehle, University of Victoria; and Dennis Connor of Dennis Connor and Associates. Also we thank Dan McCracken of the City College of New York for putting us in touch with Nick and Joan Rawlings at MUST Software International and to MUST Software International and Nick and Joan Rawlings for their generosity and support.

Part 1

Corporate Information Systems

Chapter 1

INFORMATION SYSTEMS AND BUSINESS STRATEGIES

1.1 Introduction

The main emphasis of this text is on corporate business systems. This chapter in particular deals with strategic and operational business needs for systems, and the view of information systems as models of business realities. Such insights as can be gained by the systems analyst about an organization's objectives and plans are essential if truly effective systems are to be developed; therefore these insights are extremely valuable both to the organization and to the analyst. We emphasize this point because all too often analysts are oriented toward computers and not toward business processes and objectives.

1.2 Business Strategies and Operations

A business (company, corporation or organization) can be viewed as an organized set of components composed of people, products, market, assets and facilities. In order to survive, the business must have a stated purpose or set of objectives, and it must be able to carry out these objectives in an effective manner. The objectives provide a plan of action for the entire company and criteria for aiding in making investment or management decisions. It is not adequate to merely state that the objective is just to "make money" any more than we can say that the purpose of a human being is to eat. Monetary assets are required for survival, but they are not necessarily goals in their own right.

The responsibility of setting, maintaining and implementing objectives rests with senior management. It is vital that they know the health of their company, as well as the company's potential within its own specific market area. If its particular industrial sector is growing, then the company must grow with it. If a period of decline or stagnation is imminent, then preparation for being more aggressive, being more efficient, or perhaps for entering a new area may be required. In any case, senior management are responsible for the survival of the business. They express their methods for achieving success as corporate objectives, which will be reinforced by a budget and business plan.

Typical goals in a set of corporate objectives are to:

- Increase Market Share
- Create or Enter New Markets
- Become More Efficient
- Increase Productivity
- Improve Customer Relations
- Create a Better Image
- Diversify.

These are clearly lofty phrases. In an actual business, objectives are more specific and are accompanied by a rationale and business plan. Note that objectives usually relate to the survival of the company and need to be communicated to lower management levels of the firm. (Additional information regarding objectives particular to a manager – critical success factors, or CSF's – may be found in the *Harvard Business Review* article "Chief Executives Define Their Own Data Needs" [7].)

An example of how objectives are formed comes from the airline industry. In the early 1980's, airlines faced greater competition because of deregulation, increased fuel costs, high interest rates and high labor costs. To meet these challenges, one airline's objectives included provisions for increasing market share by utilizing its reservation system in new ways, and by controlling labor costs in a novel fashion. These objectives reflected an aggressive approach to the economic climate. (Some details about the airline's specific responses are discussed in Chapters 6 and 12.)

It is one thing to set objectives and quite another to meet them. High-level objectives must be communicated to lower management levels in a company in such a fashion that they are understood and can be carried out. Here, middle management must create their own objectives and the tactics for meeting them, while always relating back to corporate objectives. The execution of middle-management objectives affects the actual operation of the company and involves all levels of employees.

In the airline example, where the reservation system was to be broadened, it fell upon the management of the marketing division to realize new marketing uses of the reservations system. In light of the airline's objectives, marketing management decided that new services should be offered to encourage travel agents to give preference to this airline's flights when available flights and connections were listed. The new services consisted of advanced seat selection, boarding pass printing, hotel and car rentals. Note the implicit need for strong support from the systems division because they ultimately implement any enhancements to the reservation system. The high labor cost problem was tackled by creating new salary levels for all new employees. This was equivalent to creating a new, parallel company. Again, systems division involvement was essential because of the need for the different classes of employees. Clearly all individuals involved in such implementations should be cognizant of the corporate objectives, for these objectives define the real purpose of the system and provide a way of measuring the system's success. Systems changes are sometimes necessary for the very survival of the company.

The administrative functions of a company – finance, payroll, purchasing, etc. – provide financial and organizational support to management and help knit the company together so that it can operate as a unit.

There are, then, three main components of an enterprise: management, operations and administration.

1.3 The Roles Of Information Systems

Information systems have been developed to meet needs in all areas of corporations. Historically the first areas were administration and operations, including:

- Payroll
- Inventory

- Accounts Payable and Receivable
- Financial Accounting
- Production
- Engineering.

The reasons that systems for these areas were accomplished first are that the areas were well defined and did not change rapidly over time, and the work was repetitive and voluminous. As computers have become more sophisticated, inexpensive and easier to use, applications have moved toward aiding management at all levels. As well, there are additional needs pushing this development: as the business environment becomes more complex and undergoes rapid changes, senior management need better quality, and faster access to, information about their own company, about their competition and about the economy. They also need means of analyzing this information and the impact of different strategies under consideration. The systems developed in this area are referred to as decision support systems. They are characterized by providing access to information and by using evaluation tools in an unstructured fashion. This means that selection criteria for data and scenarios for analysis are not known in advance, but rather are determined by the manager at the time of the request.

There is now emerging a clear indication that information systems will play an ever increasing role at the strategic level. Information systems can no longer be isolated, internal applications developed to meet individual needs, but instead must become an integral part of the company and a main delivery vehicle for products and services. For example, reservation systems in airlines were for years essential for controlling the inventory of available seats and for scheduling. Now they are becoming marketing and sales tools, and they can also deliver up-to-date information on the financial state of the company regarding bookings, on-time service and costs. We are seeing information systems emerging as powerful tools for creating competitive advantages.

A famous example is the American Hospital Supply Company (AHS), which provided on-line terminals to customers and thereby established a direct link between their customers' own order-entry and distribution systems. This resulted in an increased market share which they have maintained. The benefits to AHS's customers were the ease of ordering and the ability to track the status of an order. The benefit to AHS has been that their order-entry system is now tied to their customers' operations, making it harder for key customers to change suppliers. By being the first to offer the service, AHS has forced their competition to offer similar advantages. Many companies are now utilizing information systems as strategic tools to create new services and to help manage. In order to develop such strategic systems for an organization, it is critical that the analyst understand both the systems requirements and the business requirements. In fact, in many organizations the analyst is the one individual who has such an understanding.

1.4 The Information System As Model

The success of any information system can be measured by how well the information requirements have been captured and modeled, that is, how well the automated system reflects reality.

The software system ultimately produced is, of course, a model. It is a model in the same sense that an architectural model is a model of a building project, an investment model is a model of an investment opportunity or scenario, or an engineering model is a model of a turbine, automobile or other manufactured product. An information system attempts to model some aspect of a corporation or enterprise to a degree of accuracy such that the model is useful. Models in general are abstraction devices: they allow for the hiding of detail and for the concentration on general characteristics of what is being modeled.

At the outset in analysis, one need not be concerned with the specific functions provided by the model. Instead, one should have a more general view, and should be concerned with capturing relevant aspects of the area under investigation and ignoring others. The reason for this is that the model provides the conceptual basis for understanding, in addition to the basis for the functions contained within. The functions by themselves, however, do not form the conceptual basis.

By creating a model based on the corporate reality, the vocabulary – customers, market share, profitability, orders, products, etc. – is already defined, and is understood by all concerned. Particular functions can be described at a later time.

1.4.1 Information Systems Components

With respect to software development, any information system can be described as having two components: data and function (processes). The functions will ultimately be expressed as software modules which may produce summary financial information, project future profits, provide an analysis of sales by product line within region and so on. The functions can provide adequate information only if the appropriate data are present and can be extracted. A system, then, has these two separate components that are, in fact, *duals* of one another. Of the two, data form the foundation of the system, for without the data, the functions could not exist. It is likely that the data will eventually take the form of an automated, commercial database management system (or part thereof), as a set of stand-alone computer files, or perhaps as some combination of both. However, the analyst is concerned only with understanding, defining and specifying the user's data requirements, regardless of the form they will eventually take.

1.4.1.1 Data

Data are a resource for an enterprise, and of late, much press has been devoted to topics such as "Information Resource Management," "Exploiting the Data Resource" and the like. The example of American Hospital Supply discussed above illustrates such exploitation of the data resource.

In fact, studies at MIT's Sloan School of Management have suggested that strategic improvements in information systems uses are coming from *outside* the systems development area, especially in areas where there is customer contact. Such systems are rooted in traditional systems development. The innovative use of the data is what distinguishes such systems. As well, the same studies show that there is at least as much potential for information systems development in these areas as from those areas that have already undergone automation. These strategic improvements have, to a large extent, been motivated by the shift in market place from a seller's market to a buyer's market and have been facilitated by advances in technologies and changes in approaches to software development.

Some enterprises may not exploit this resource; however, it still exists, whether used effectively or not. An information system, then, is a vehicle for the exploitation of data – one expects to distill useful information from it. Data provide the basis for the system, and for this resource to be used in applications development projects, it must somehow be represented in a way that is amenable to automation.

Often these two aspects are conflicting: the analyst must represent or organize the potential user's data requirements so that they are at once representative of the user's world, but at the same time conducive to automation. (In Chapter 3, the characteristics of data are discussed and a commonly used data modeling tool that accomplishes both tasks is explored.)

1.4.1.2 Functions

The functions or processes within an information system are the dual of the data component. They are the software modules that extract, manipulate, and generally manage the data. The functions, and the information system as a whole, can be of use only if the relevant data have been captured.

Assuming that the appropriate data have been modeled, the functions provided by the system must satisfy the needs and expectations of its users, and to that end are modeled during systems analysis. The business functions will eventually take the form of code in a particular programming language, and the job of the analyst is to

- understand what functions the system must perform
- model the functions in a way favorably disposed to automation.

The particular method of function modeling chosen by the authors is discussed in Chapters 7 and 8; it enjoys widespread use throughout North America, Europe and Australia. The process of modeling the functions endeavors to capture the business processes that must be automated. Of these two components (data and function), data provide a more stable and global foundation for systems development, for if the functions cannot provide the required information, then the appropriate data must not have been captured. Functions, however, are local to particular systems projects. An alternative way of stating the relationship between function and data is that only if the pertinent data have been modeled can the required business functions be automated.

1.5 Exercises

1. Why does a corporation need specifically stated objectives?

2. What objectives might a small company have?

3. How can information systems be utilized as part of a strategy?

4. If you write out your monthly budget on paper, what is being modeled?

5. It has been stated that data are a resource, and as such have value (not necessarily monetary). Of what value are your own monthly cash flow data to you?

1.6 Bibliography

1. Cash, James I., Jr., McFarlan, F. Warren and McKenney, James L. *Corporate Information Systems Management: Text and Cases*. Homewood, Illinois. Richard D. Irwin, 1983.

2. Dampney, C.N.G. et al. *Data (Information) Analysis of Business Systems - Informal Notes*. Waterloo, Ontario. University of Waterloo, 1986.

3. Gray, Peter. *Logic, Algebra and Databases*. Chichester, U.K. Ellis Horwood, 1984.

4. Jackson, M.A. *System Development*. Englewood Cliffs, New Jersey. Prentice-Hall, 1983.

5. McFarlan, F. Warren and McKenny, James L. *Corporate Information Systems Management: The Issues Facing Senior Executives*. Homewood, Illinois. Richard D. Irwin, 1983.

6. Mosevich, J.W and McCandless, W. *Managing Fourth Generation Technologies: Guidelines for Action. CIPS/ACI Congress 86*. Vancouver, British Columbia, 1986.

7. Rockart, John F. *Chief Executives Define Their Own Data Needs*. Harvard Business Review, 57, 2, March - April, 1979, pp. 81-91.

8. Rockart, John F. and Bullen, Christine V. (editors). *The Rise of Managerial Computing: the Best of the Center for Information Systems Research*. Homewood, Illinois. Dow Jones-Irwin, 1986.

9. Tsichritzis, Dionysios and Lochovsky, Frederick H. *Data Models*. Englewood Cliffs, New Jersey. Prentice-Hall, 1982.

10. Weinberg, Gerald M. *Rethinking Systems Analysis and Design*. Boston, Massachusetts. Little, Brown and Company, 1982.

11. Zachman, J.A. *A Framework For Information Systems Architecture*. Los Angeles, California. IBM Los Angeles Scientific Center. March, 1986.

Chapter 2

THE SYSTEMS DEVELOPMENT LIFE CYCLE

Information systems analysis is concerned with the conceptual design of software systems and environments for meeting the needs of businesses and organizations. The process of systems analysis is accomplished by gaining a thorough understanding of the business needs through the analysis and modeling of relevent information requirements and processes. The end result is a specification or "blueprint" for developing solutions to the business requirements. As discussed in this book, systems analysis is the most important part of a larger task, the process of software development, called the *systems development life cycle.*

The term *software* refers to sets of computer instructions, and the term *systems* refers to the integrated set of components of software modules, databases, hardware modules and those non-computer entities relevant to the business needs being met.

The activity of systems analysis is important and challenging for a variety of reasons. Firstly, systems analysts deal with a variety of people with pressing business requirements that are often vaguely understood. Yet they must produce conceptual designs of software systems that are well-defined and structured. Secondly, even if a business situation is well understood, there is never just one automation solution but many alternative options. In fact, automation may at times not really be *required*, but rather just *desired.*

In any case, software systems are the ultimate result of systems analysis. It is helpful, then, to consider the different types of computer systems currently in use and also to view software generally in terms of its attributes and how they relate to systems analysis.

2.1 Software

The fruits of the systems analyst's labor will eventually be transformed into some form of computer code, designed to run on a certain hardware configuration, according to specified procedures. On the one hand we have a concept in the mind of the systems analyst and the customer (the "user"), and on the other hand we will have the reality of some final product. The measure of success in creating an acceptable final product is almost always the software since it is the logical and physical expression of the concept. One should keep in mind that even with well-understood requirements and concepts, it is not a trivial task to create software that really does match expectations.

"Building" software is different from almost any other constructive activity. If a simply stated problem is given to a class of thirty students, one will get thirty distinct programs in return. Often identical input will result in different output; some of the programs will not run under certain conditions and some not at all. Since this can happen with simple problems, one can imagine many possibilities with large corporate systems. Thus programming and software design is currently an individual activity. Analogies between building software and building cars or airplanes are not really valid except in limited ways. Creating software is a unique, challenging endeavor.

A second point about software is that it is difficult to know if it is "correct." The concept of correctness is interesting. Does it really mean anything to say that a particular program is correct? Why, if it meets requirements and performs as expected, of course! In other words, for a particular input, we get the output required by our concept of what the software is supposed to do. Thus, one can never divorce the systems analyst's and user's concept from the end product's performance. If one can write down all possible inputs and conditions with corresponding outputs and conditions, and if these are few in number, then one might be able to test a program for correctness. Obviously, in most applications this is impossible. We must currently live with the fact that while we try our best to design and test software, we cannot guarantee its total correctness.

Another feature of software to consider is its quality. If no two programmers produce exactly the same code, and if we can rarely guarantee correctness, can we conceive of software in terms of quality? The answer is a qualified yes. We can conceive of quality, but again it is difficult to judge a specific piece of code for quality. However, we can *aim* for quality.

What do we attribute to quality software and what do we aim for? Firstly, even if we cannot prove correctness, if we and others can at least understand the code then we have some hope of its being correct and of fixing it if required. Thus the first property of quality is understandability.

Secondly, if a program needs to be changed or enhanced to meet new requirements for its user, it must be possible to effect such changes and also to trace them, because they will undoubtedly affect other programs. Quality implies maintainability.

There is another attribute of quality software that has to do with how well it meets the requirements and needs of its users. If users are continually requesting enhancements and changes to a program, or are not using it because it does not perform as required, then one would probably judge the program's quality as being low. This points again to the link between the systems analyst's concept, which reflects requirements, and the coded product. Quality software, then, is usable and meets users' expectations.

Ideally, software must be of high quality, and must meet the user's business needs, the primary reason for its creation.

Finally, software is not static – in fact, it is never really completed. Even if it is correct when first turned over to users, there will almost always be changes required. Most common changes occur because of new or different user requirements. For example, if your competitor's system suddenly has a new feature, you will probably need a similar or better one for your business to survive in the marketplace. Sometimes, after a few months of using a system, a user will realize limitations or think of valuable new features that could be incorporated. Often the external environment changes (for example, a change in federal tax regulations) and the software needs appropriate modification. Other areas of change include the operating system and hardware environments. Often users outgrow their systems or systems are so outmoded they are scrapped.

In any case, the systems analyst must never view software as being static, and ideally should design accordingly by making it understandable, easily maintainable, flexible and correct.

2.2 Systems Perspectives

The emphasis of systems analysis is on understanding a business situation and its environment. There are several possible systems environments in which automation for the business situation will be the most appropriate choice. These different environments can be viewed from several perspectives and should be kept in mind during analysis. One caution, however: analysis is concerned with the application first, the system later (the system is a function of the application). Thus, one must not let a system define the application – "Form Ever Follows Function" [6]. Analysis defines the function, and later activities in the software development process give the form.

Following are some different views and perspectives of systems.

Functional Perspectives

Systems can be viewed from the perspectives of what they do and why they exist.

Operations/Production Inventory, Reservations, Automatic Teller Machines.

Administration Payroll, Corporate Budget, Accounts Payable and Receivable, Office Automation.

Information Retrieval Library Search, Management Information.

Decision Support Statistical Analysis, Financial Modeling, Simulation.

Design/Engineering Computer-aided Design.

Manufacturing Computer-aided Manufacturing, Material Requirements Planning, Robotics.

Personal Computing Spreadsheet, Word Processing, Personal Filing.

Processing Mode

There are several modes of operation of systems that affect responsiveness, costs and systems attributes as a whole.

Batch Programs are queued and executed according to a schedule. This can take place almost immediately or hours after submission. There is no user intervention.

Interactive Programs are executed on-line with user interaction being required. File updating may not be instantaneous.

Real Time Files are updated immediately and almost simultaneously.

Hardware Environment

There are many options regarding the type of equipment on which systems can be implemented.

Centralized/Mainframe Large computers that can support hundreds of simultaneous users and programs. Usually cost millions of dollars.

Minicomputer Medium capacity computers that can support several simultaneous users and programs. Often dedicated to one type of application. Cost from fifty to five hundred thousand dollars.

Personal Computer Microcomputer for one person at a time. Costs in the hundreds or low thousands.

Distributed Systems It is possible to link computers of different types.

Communication Environment

Direct Connect Terminals are wired directly to the computer.

Dial-in Connection is over phone lines.

Local Area Networks Workstations are all connected on one network. Data, devices and software are shared.

Wide Area Networks Computers and terminals are connected over long distances.

Software Development Environment

Traditional Procedural code such as FORTRAN or COBOL is used in the traditional setting where software is written in-house.

Packages Programs are purchased, modified if necessary.

End User Fourth-generation languages are used by non-programmers.

Information Centers Corporate information is provided on computers.

Prototyping Users and analysts together develop systems iteratively.

The above are not meant to be a catalog or checklist, but rather are meant to present in simplified form some of the attributes to consider when conceiving a system designed to meet business needs.

Some applications lend themselves naturally to one type of system, while others must be judged on both immediate needs and potential growth or change. For example, if a requirement is for information that can be a day or more old and for which the user can wait several hours then it is probably valid to think in terms of a batch environment. If the need is for fast answers, but with data still a day or more old, then perhaps an interactive approach is required. And finally, if information must be precisely up to date, as in a reservation system, then a real-time mode is called for. The exact requirements regarding timeliness and response must always be determined during analysis.

2.3 An Overview of Analysis

Systems analysis is concerned with the inception and conceptual design of software. Be it a small personal computing application or a large corporate real-time system, software is essentially the implementation on a computer of some logically definable procedures. Underlying and defining these logical procedures is a purpose or objective of the software, such as balancing and keeping track of a person's budget or controlling a manufacturing company's inventory. Regardless of how small or large a software product is, it is essential that the underlying processes, procedures or objectives be understood in sufficient detail so that the finished software system does exactly what is required. The translation from "needs" to final software implementation can take a few minutes, as in the case of a simple budget on an electronic spreadsheet, or it can take months of team effort, as in the case of a real-time banking system. Systems analysis is concerned primarily with the process of gaining an understanding of the business needs, translating them to form a logical model, and conceiving a system to improve or support the business area in question.

As we all know, computers are becoming cheaper, more powerful, easier to use, and more pervasive in our lives. But beyond this, computers, and more precisely software, are now essential for the survival of most major corporations, and even for the protection of entire nations. This should impress upon data processing students and practitioners alike, the importance of high quality systems analysis and design, for it is here, especially at the analysis phase, that mistakes are really costly. A faulty design can be repaired only by "going back to the drawing board" (if there still is a drawing board!). In addition, the very precision of software makes it difficult to change. So if it does not perform as its user specifies or expects, it may not be of much use.

Despite the fact that all software systems start with systems analysis, it turns out that different degrees and types of analytical techniques exist, depending both on the application and on the corporation. If you are using a spreadsheet on a personal computer for balancing your checkbook, the analysis can be done mentally or by experiment. If, however, the application uses many spreadsheet cells, formulae, graphs and files, often preliminary work must be done on paper, showing flow of data, file input and output, and report formats. At the other end of the spectrum, if the system will be implemented on a large corporate mainframe where dozens of new and existing files, programs and procedures will be involved, it is essential that a precise model be created showing all processes, data flows and files, human-machine interfaces, data communication requirements and so on. Further, in this case there are usually many new developments occurring simultaneously in a corporation. Thus there is a need for a standard analysis and design methodology, not only to impose some sort of uniformity, but also to ensure compatibility with the existing systems environment.

As alluded to above, there is a great diversity of types of applications and systems environments. The fast pace of evolution in computer hardware and software certainly has an impact on systems analysis; indeed it should be taken advantage of. Such developments as personal computers, local area networks, fourth-generation languages, artificial intelligence, automated systems design tools and application generators are a few that come to mind. These impact traditional systems analysis by providing capabilities not possible with earlier generation software such as COBOL and FORTRAN. Indeed, determining requirements (getting at what the user wants) can be greatly improved by prototyping a system using some of the new tools. Packaged software avoids programming altogether. Recently, automated systems design tools and application generators have been developed which produce source code directly from requirements, thus eliminating programming. This is a very promising and exciting development. Naturally, there are certain tradeoffs with these systems and while we have not yet reached the stage of eliminating COBOL, FORTRAN and programmers, they have already made a great impact on systems development, and their impact is expected to, and in fact must, increase.

So, regardless of the approach taken, before code is written or the personal computer switched on, we must understand what is to be accomplished, first from the business viewpoint and later from the computer viewpoint. The process of transformation from business needs to logical model is systems analysis.

2.4 Traditional Software Development

Most software development in medium to large companies occurs in organized projects according to a process referred to as the traditional life cycle (sometimes called the life cycle, systems development life cycle (SDLC) or just the traditional approach). Basically, this is a series of well-defined steps starting with the analysis and specification of user needs and ending with the implementation of a live system. It is performed by a team, sequentially, from one phase to the next. We refer to it as traditional because there are new approaches emerging for software production, covered in Part 3 of this text.

2.4.1 Background

The commercial data processing industry is only about thirty years old. Early computers were programmed in very low-level languages like assembler, and later in languages such as COBOL, FORTRAN and PL/1. Applications were initially fairly simple and over time have evolved to become more and more complex. Eventually a point was reached at which crises started occurring in the industry, the symptoms of which were late projects, applications that did not perform as expected (by the user), unmaintainable, patched-up programs, out-of-date or absent documentation, systems incompatible with current procedures and so on. As "necessity is the mother of invention," many corporations, and industry in general, developed methodologies in an attempt to "engineer" software, that is, to create a process for its inception, development and implementation. These methodologies as a group are referred to as the traditional life cycle (TLC).

Because it was developed within a span of only ten years during the 1970's, it deserves to be called a monumental achievement. However, because it was designed essentially for so-called second- and third-generation hardware and software (assembler, FORTRAN, COBOL, etc.), and because it is predominantly a manual process (there are few automated aids), it has a great number of faults and limitations. Nevertheless the TLC will be required for quite a few more years – not just because COBOL and FORTRAN will be used for a long time, but because it contains certain attributes and functions generally required for systems inception, development and implementation, regardless of techniques used.

Specifically, any major software development in a corporation is an investment, and as such, requires management's understanding and approval. As will be shown, the TLC has certain milestones at which management approvals are natural. Furthermore, the life cycle is built upon the process of creating and developing in a precise, systematic and well-thought-out manner. This is because current second- and third-generation programming languages require that procedures be defined to a primitive level. The effort required to think things through, especially in the initial phases of the cycle, is beneficial both for designing a system and for the business as a whole.

Part 3 of this text explores the major problems with the traditional approach and how they are being attacked, but for now we shall present a generic life cycle, because

1. it is common in most companies, and

2. it provides an excellent introduction to the topic of systems analysis.

Most companies have a corporate life cycle process that they either subscribe to (an existing commercially available methodology) or that has been developed in-house. The authors present here a model for the TLC that should provide students with a basic understanding of the software development process that will serve them in the business environment. It is divided into six phases:

- Requirements Determination
- Systems Analysis
- Systems Design
- Programming
- Testing
- Implementation (see Figure 1).

For each phase, its purpose, procedures, management decision points and end products are described. There are many other such models, some with fewer or more phases, but all incorporating the same basic procedures and principles.

The purpose of the TLC is to provide a method for producing quality software, as described earlier in this chapter. What is meant by quality from the computer programmer's viewpoint? This may mean using good structured programming constructs, choosing meaningful variable names, writing modular programs and so on. To the designer, quality software may mean a factored system: a system composed of modules, each of which performs one function and is relatively independent from other modules.

Both of these views are too narrow and neglect the fundamental fact that software must always be judged from the corporate viewpoint. It should be viewed as something that produces a result, requires servicing, incurs ongoing costs and provides a benefit. The result that is produced must be cost effective, and it must be worth the effort (in real dollars or perceived value) in order to be of some worth to the organization.

It has been estimated that maintenance accounts for 60 to 80 percent of the total cost of an average system, and the average life of a system is only five to seven years. Given these statistics, the motivation for producing a quality product is strong, and to ensure a quality product, the TLC is the method by which systems development has been approached.

2.4.2 The Life Cycle

The development process described here consists of six phases which occur in chronological sequence. Naturally, as work progresses in any one phase, some cycling back will be required because of unforeseen problems or misunderstandings between systems developers and users.

2.4.2.1 Requirements Determination

This first step in the software development process is a very high-level assessment of a software development request.

Input to This Phase: A formal or informal request from a user or department to data processing or the systems development group identifying business needs.

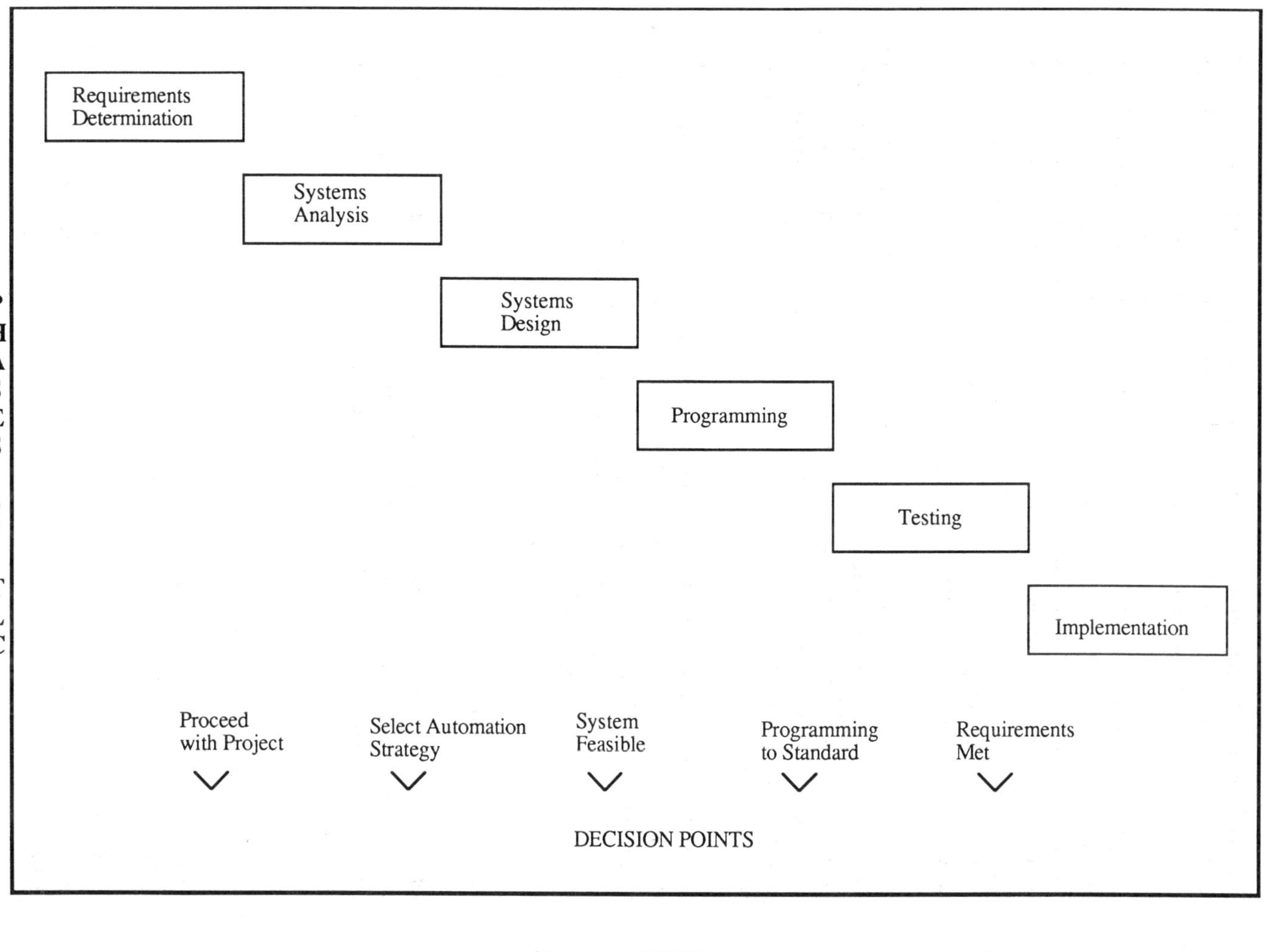

Figure 1: Management decision points in traditional software development

Output from This Phase: Requirements specification document.

Management Decisions: That there are or are not sufficient benefits, given rough cost and technical constraints, to proceed with systems analysis.

Before the systems development personnel receive a request that initiates a study of user requirements, there may be considerable activity in the user area. Obviously, there are business needs not being met by current procedures. These needs may take the form of informational needs, for example, management may perceive a need for information regarding sales by product for each salesperson, information that is either unobtainable or not easily obtainable under current circumstances.

The need may be of an operational support nature, for example, many companies physically count their inventories of stock at the end of accounting periods, such as at the end of a year's performance, or perhaps semi-annually or quarterly. Often when performing a physical count, the business must cease operations (sales) to ensure an accurate accounting. These are often viewed as lost days. An alternative, called cycle counting, may be preferable. With cycle counting, counts of certain items are performed periodically, eventually rotating through all stock. However, cycle counting may not be possible currently because of existing procedures or lack of system support. Whatever the nature of the business needs, they are formulated within the affected business area and surface as a request to the systems group.

It has been estimated that there is an average two- to three-year backlog of software development projects within many corporations. Systems personnel generally receive many more requests for projects than they can handle.

Within this first phase, which begins after a request from a (potential) user has been received, three activities take place. First, the problem or need is identified. Often users will phrase their needs in terms of a solution, i.e., "we need a new payroll system that will automatically interface with the general ledger," or "our inventory control system should be capable of giving turnover rates per year." Depending upon the size and nature of the project, this may be a small, relatively easy task for an analyst, or it may involve several analysts working with users for large, more complicated projects.

Second, after the needs have been identified, the request must be evaluated. In many businesses the evaluation may involve committee participation. The economic (cost versus benefit), technical and operational feasibility of the project is roughly assessed.

Finally, pending management approval/rejection of the request, the results are documented. If the project has been approved, the results of this phase, the requirements specification document, eventually becomes the mandate for the analyst to begin work, and it establishes general terms of reference within which the analyst will operate.

2.4.2.2 Systems Analysis

In the second phase of software development, a logical model that will satisfy the user's business requirement is conceived.

Input To This Phase: Requirements specification document.

Output From This Phase: Functional specification.

Management Decisions: The functional specification is a logical model of the conceived system, together with estimates of the associated costs and benefits. Given this, management must decide whether to proceed further with the development of the potential project as is, to modify or to abandon the project.

Systems analysis is the most creative phase of the software development process. The analyst must develop an understanding of the business requirements, which are normally expressed in business terms, and transform them into a logical concept in systems terms that is capable of meeting those business needs. There are obviously many different solutions possible, some of a standard nature and others totally novel. We divide the analysis phase into six steps, which are discussed in more detail in Chapter 6. These six steps deal with the following:

1. studying the user's environment, and producing a physical (procedural) model

2. abstracting the physical model into a conceptual or logical model of the present system, be it automated or not

3. using the conceptual model of the present system as a basis, developing a logical model of a system to meet the user's needs

4. developing automative options or possibilities and supporting documentation including cost/benefit analyses, preliminary budgets, work plans and schedules

5. selection by management of an option, and finally,

6. packaging the selected system into the main product of analysis, the functional specification.

In addition to data and functional concepts, hardware requirements are estimated in this phase and included in the functional specification. This may include a recommendation for the purchase of new equipment or expansion of existing equipment. If management approve continuation into the next phase, design, the hardware decision is finalized and will be an important consideration for the designer.

Although not usually considered part of analysis, but rather a separate exercise, there is opportunity in this phase for the user-analyst team to create or change existing business processes – both for improvement in their own right and for getting the most out of automation. One reason for this opportunity is the effort put into understanding the business area and the identified need for change. Collectively, the user and systems developer exhibit a synergy that allows for optimizing business processes and the potential benefit of automation.

One possible result of systems analysis might very well be a recommendation *not* to proceed with the project. This may be because of insufficient benefits, infeasibility of current technology, lack of resources or other projects having higher priority. Thus one must never assume that automation will result just because of a need.

2.4.2.3 Systems Design

In the third step of the traditional life cycle, the logical design of the software is translated to a physical design.

Input To This Phase: Functional specification.

Output From This Phase: Systems design document (systems blueprint).

Management Decisions: Does the design adequately match the functional specification? Will the end product provide adequate benefit?

In the design phase, the new system begins to take form. The functional specification (sometimes called the system specification) states what the system must do to meet the user's needs, and is the product of the analysis phase. It shows *what* is to be done, while the product of design formulates *how* it is to be accomplished. The design process consists of incorporating the logical design from analysis within the existing (or new) physical environment. One can view the end product, the systems design document (sometimes referred to as the design specification), as a blueprint of the system conceived during analysis.

The activities that typically occur during the design phase include the following.

1. The *outputs* required by the customer are completely specified. This may involve the design of screens in interactive systems, layouts of hard copy reports and specifications for special forms such as payroll checks, purchase orders, share certificates and customer billing statements. Sample reports are developed and the frequencies, media, destination and use of outputs are described in detail.

2. Because the outputs determine the *inputs* that are required by the system, the volumes, sources, media and documents regarding the inputs necessary are specified.

3. The *physical file system* is derived. The organization, content and access paths are decided upon and fully documented. The system hardware and software capabilities play a major role in determining the design of the physical file system.

4. There are two levels of *processing* that should be considered: the higher level of overall systems processing, and the lower one of processing at the program level. The former deals with the whole system at a broad level and may include the development of systems flowcharts which show the flow of control among program modules within the systems as well as the interfaces with other systems, security and controls considerations, and implementation plans. Lower-level processing involves documenting all program module specifications: the inputs, outputs, and the processes performed by each module as well as the programs necessary for conversion to the new system.

5. The systems design document also includes descriptions of manual procedures surrounding the automated system and for system backup and recovery.

2.4.2.4 Programming

Input To This Phase: System design document.

Output From This Phase: Computer code, user manual, operator manual.

Management Decisions: To date, does the system match expectations and requirements? Is the timetable for the development period being met? Does the system being developed have any major deficiencies?

In the programming phase, the code for the system is produced along with the user documentation: this describes the system from the user's perspective and provides instructions for its use. As well, instructions for the computer operations staff may be developed during this phase, along with a testing plan.

2.4.2.5 Testing

Input To This Phase: Coded programs, documentation, and testing plan.

Output From This Phase: Test results, change requirements, implementation plan.

Management Decision: Does this system meet requirements?

The purpose of the testing phase of software development is *not* to exhibit the absence of errors in the programs being tested, but rather to find as many errors as possible [14]. This statement may seem to contradict one's intuitive notion of a successful test of a piece of code; however, a successful test is one in which errors have been found, not one in which no errors have been detected.

In this step, usually two methods of testing are employed: human testing and machine testing.

1. Human Testing
   In most companies where systems are developed, the first testing that occurs is not machine testing, but rather the detection of errors through manual, team efforts referred to as walkthroughs and inspections. With these methods, the author of the program forms a group with others who represent programming expertise or who may be part of the project team.

Manual error detection has become increasingly important and has proven to be an effective method of uncovering faults. A fundamental premise in software development is that the earlier mistakes are detected, the less costly they are to correct and the more likely they are to be corrected accurately. Humans are capable of pinpointing the error itself, while a machine can find only the symptoms of the error.

In a walkthrough, a group (usually consisting of members of the current software development project) develops test cases and "plays computer" by stepping through the logic of the program. Program inspections involve the author of a program reading code aloud to an audience. Both methods have been shown to uncover 30 to 70 percent of the errors in logic and coding, and both are common practice in most corporations.

Fundamental to both methods is the belief that one cannot competently find errors in one's own work. In fact, it is likely that if we were each responsible for detecting our own errors in programming, we would likely contrive biased test cases (probably unconsiously) which would exhibit the lack of, rather than the presence of, errors. The term "egoless programming" was coined by G.M. Weinberg [19]. It refers to the ability to deal with the reality that one's own work contains flaws, and also refers to the methods of walkthroughs and program inspections used in error detection. In both walkthroughs and program inspections, test cases are chosen to detect the most errors.

2. Machine Testing
After code has been subjected to the manual testing described above, it is then machine tested. Again, the test cases are designed to detect the maximum number of errors.

Unit or module testing refers to the process of testing individual subprograms, modules, subroutines and procedures within a program. The purpose of unit testing is to compare the system's functioning to its specification as defined in the systems design document.

Incremental or integration testing occurs subsequent to unit testing. This means combining modules with those that have already been unit tested. The incremental approach not only tests the logic of each module, but also the "rippling effects" of errors, and the interfaces among modules.

Eventually each module within the system has been incorporated, and incremental testing is complete. At the same time, interfaces with other systems are also tested via incremental testing.

3. Higher-Level Testing
Although inspections and walkthroughs are effective in detecting coding and design errors, and machine testing can uncover further mistakes, neither are appropriate for detecting higher-level errors in requirements and analysis. That is, do the programs produce what is required by the user?

If a system or program does not do what its user expects it to do, then of course there is an error in the software. Software development, to a great extent, is a process of communicating information, beginning with the user's individual requirements. After the requirements have been modeled during analysis, they are translated into another form during design, and into yet another form during programming. (Software development surely represents a tautology.) Many software errors can be attributed to mistakes, incomprehension and breakdowns in communication and in the translation processes [3]. The ultimate testing of software is not to compare its functioning to the systems design document or to the functional specification, but to compare it to the original objectives, the expectations and the current needs of its users.

The testing phase of the development process is complete when the user accepts that the system really does meet the determined requirements.

2.4.2.6 Implementation

Input To This Phase: Implementation plan and the tested system

Output From This Phase: Live system

Management Decision: To fully utilize the system or to revisit a previous phase.

Implementation involves the activities necessary in converting to the new system. During this phase, computer center personnel are trained in the operation of the system. Users are also trained. During implementation, the final conversion to the new system takes place, and it can occur in various ways: both systems may be run in *parallel* for a period of time, there may be a *direct cutover* from the old to the new system, or a new system may be implemented using a *phased-in* approach [17]. Regardless of how the actual conversion is done, a live, fully operational production system is the final product in the software development process.

2.5 The Software Development Project

Software development projects may originate from any area of a corporation, may pertain to different business applications, and may have various objectives. All the objectives have the following in common: a customer or user of information has a business need that is not met by current procedures and practices, be they automated or not. The need may be perceived as internal to the company, for example providing information more rapidly, or accurately, or providing new information. The need may be perceived as arising from factors external to the organization, such as a change in government regulations. From whatever source and for whatever reason, all projects originate with a "need."

The composition of the project team varies depending on the particular company and its internal organization. Normally, there is heavy involvement of the user at the beginning of the project. This can range from full time to part time, and from one to several people. Users may be from different levels within the organization. Usually a "responsible" user from middle management will be involved, as well as users who will utilize the system directly. Almost always, the same individuals from the user community are involved from start to finish. Project leadership also varies and can be from either data processing or from the user group, but if it is a user in the beginning, it will probably change to data processing during design.

Involvement in the project from the data processing department also varies in numbers, depending on the magnitude of the effort. However, the specific people involved invariably change as the project progresses because different systems skills are needed for each phase of the project. At the beginning, there will normally be one or perhaps a few systems analysts on the project, with occasional input from other skill areas as needed. Later, in the design phase, there will be involvement of systems designers, database designers and, of course, the systems analysts. After the design phase, the people most involved will be programmers. So in the traditional approach, the composition of a project often changes from one phase to the next.

2.6 Exercises

1. Draw an analogy between creating software and designing and manufacturing an airplane or car. Relate this to the traditional life cycle.

2. With respect to Question 1, explain why a change in requirements is more expensive later in the project and can even be disastrous. (Relate this to airplane or car design.)

3. What are the benefits of the life cycle?

4. What are the drawbacks of the life cycle?

5. If computer programs could be generated by computers themselves, would analysis still be required? Why?

6. For as many computers or computer systems as you have had contact with, classify them according to the perspectives given in this chapter.

7. Just how important is software? List as many examples as you can of life-saving, defensive or critical software (e.g., in aircraft, medicine, military, etc.).

8. Assuming that brains are computers, explain why dinosaurs are a good example of great hardware but poor software. What about humans?

9. Why is analysis so important?

10. What problem with today's software can cause the cost of testing programs to represent 50 percent of the total cost of software development?

2.7 Bibliography

1. Biggs, Charles L., Birks, Evan G. and Atkins, William. *Managing the Systems Development Process*. Englewood Cliffs, New Jersey. Touche Ross and Company, Prentice-Hall, 1980.

2. Boar, Bernard H. *Application Prototyping: a Requirements Definition Strategy for the '80s*. New York, New York. John Wiley & Sons, 1984.

3. Brooks, Fred P., Jr. *The Mythical Man-Month: Essays on Software Engineering*. Reading, Massachusetts. Addison-Wesley, 1975.

4. Brown, David B. and Herbanek, Jeffrey A. *Systems Analysis for Applications Software Design*. Oakland, California. Holden-Day, Inc., 1984.

5. Curtis, Bill (editor). *Tutorial: Human Factors in Software Development*. Los Angeles, California. IEEE Computer Society, 1981.

6. DeMarco, Tom. *Structured Analysis and System Specification*. Englewood Cliffs, New Jersey. Prentice-Hall, 1979.

7. DeMasi, Donald. *An Introduction to Business Systems Analysis*. Reading, Massachusetts. Addison-Wesley, 1969.

8. Gane, Chris and Sarson, Trish. *Structured Systems Analysis: Tools and Techniques*. Englewood Cliffs, New Jersey. Prentice-Hall, 1979.

9. Gildersleeve, Thomas R. *Successful Data Processing System Analysis, Second Edition*. Englewood Cliffs, New Jersey. Prentice-Hall, 1985.

10. Hodge, Bartow and Clements, James. *Business Systems Analysis*. Englewood Cliffs, New Jersey. Reston Publishing, 1986.

11. Jackson, M.A. *Principles of Program Design*. New York, New York. Academic Press, 1975.

12. Jackson, M.A. *System Development*. Englewood Cliffs, New Jersey. Prentice-Hall, 1983.

13. Martin, James. *An Information Systems Manifesto*. Englewood Cliffs, New Jersey. Prentice-Hall, 1984.

14. Myers, Glenford J. *The Art of Software Testing*. New York, New York. Wiley Interscience, 1979.

15. Parkin, A. *Systems Analysis*. London, U.K. Edward Arnold, 1980.

16. Pressman, Roger S. *Software Engineering: A Practitioner's Approach*. New York, New York. McGraw-Hill, 1982.

17. Senn, James A. *Analysis and Design of Information Systems.* New York, New York. McGraw-Hill, 1984.

18. Weinberg, Gerald M. *Rethinking Systems Analysis and Design.* Boston, Massachusetts. Little, Brown and Company, 1982.

19. Weinberg, G.M. *The Psychology of Computer Programming.* New York, New York. Van Nostrand Reinhold, 1971.

Chapter 3

THE CORPORATE DATA MODEL

3.1 Introduction

The rise in the importance of data within the corporation together with the advances in database technology has caused the management of data to become an organizational goal. If corporate data collectively were to be modeled around the subjects of interest (i.e., customers, products, and the like) then a more unified view could result, which would be more stable and suitable to everyone than before. Such a view of data would have to proceed from a top-level perspective in order to capture what is important to the enterprise. This chapter explores data and data modeling, and discusses a modeling tool that has become widely used throughout North America, Europe and Australia.

3.2 Data

One can define data as descriptions of events [18]. They correspond to discrete, recorded facts about events from which one obtains information. *Information* is an increment of knowledge inferable from data.

Data have both a factual representation and a connotative meaning or interpretation. For example, "the temperature today is 30 degrees Celsius" has the representation within the quotation marks, as well as the meaning that today is a hot day. In automated systems, the data and their meanings are separated. The computer can process the facts or data, but their interpretation is left to the users. For example, on Friday, September 5, 1986 the New York Stock Exchange experienced its largest loss since the stock market crash of

1929. Afterward, financial analysts attributed partial responsibility to the Exchange's new computerized information system which provided information regarding selling trends as they were occurring. Early in the day the market started to fall, and as the users (traders, stockbrokers, etc.) became aware of the situation, more and more selling occurred, thereby exacerbating the trend. A more mature interpretation (or more correctly, exploitation) of the data might have been to stop selling, in order to reverse the trend. Perhaps in this situation the users believed that the meaning lay with the data.

3.3 Data Modeling

A data model is an abstraction device, that is, it provides a way to concentrate on the meaning or information content of the data and to disregard the detail, as in the old cliche "to see the forest as opposed to the trees" [24]. Not only does the data model relieve one from a lot of detail, but more formally, it provides the conceptual tools for describing data and data relationships, for understanding the semantics (meaning) and for grasping the constraints (rules) regarding data operations. One presumes also that the actual data provide a relatively stable foundation for modeling an enterprise, while at the same time capturing the dynamic aspects of the world which will be automated.

Data models are used in many fields: finance, statistics and civil and electrical engineering to mention a few. The kind of data modeling that is of interest within the context of information systems is based on the mathematical notion of a set. The concept of sets is fundamental to mathematics and to all mathematical applications. A set is a group or category of objects, together with a membership condition. For example, the set of students of a college may be defined as those people currently enrolled in at least one course. The set of sales personnel of a particular company are those employees who sell the company's products. To describe a set using a membership condition, or by definition, is to describe it by *intention.*

A set may also be described by naming its members. For example,

$$S = \{1,2,3\}$$

describes the set S with three members, the numbers 1, 2 and 3. The order of the members or elements is immaterial. To describe a set by specifying the members is to describe it by *extension.* The elements of a set are presumed to be distinct, so even though there may be duplicates, only one occurrence of each member is written. The set {2,3,4} is the same as {2,2,3,3,3,4,4,4,4}.

Finally, a set can have a finite or infinite number of members or elements. For example, the set of positive integers is an infinite set, but the set of products offered by a company is finite.

In information systems, one wishes to concentrate on general common properties of sets of objects in order to obtain categories or *types* of data. The objective of such effort is to produce a global data model – a unified view – which can provide the data (database) framework for all information systems. The information or data model provides the foundation of the conceptual database (database here means the *working* or *operational* data). Much of database theory and knowledge is closely related to mathematical set theory, and to mathematics generally. It is important, then, to derive a "good" database – one that satisfies users' requirements and that behaves well under automation. The authors' treatment of data is directed toward this objective.

The kinds of data of interest to a user and an analyst for any specific software development project are just those of interest generally to the enterprise: customers, products, orders, suppliers and so on. Such categories of objects of interest in an enterprise are referred to as *entity sets* or *entity types* in database design. An entity is some discernable or distinguishable aspect of the world.

When considered as a whole, the data in an enterprise can be thought of as being of one type or another. Each particular type of data will belong to a set that includes all other data of the same type. For example, one can speak of the entity set PRODUCTS, the entity set CUSTOMERS or the entity set SHAREHOLDERS. Within each entity set, for example CUSTOMERS, the individual members are referred to as entities. Russ Jones, 123 Main Street, Trotwood, is a customer and is distinguishable from other customers. In information systems, what is attempted first, is to model these data abstractions globally, in a way conducive to automation, and *then* to build local models of the required functions. (*Local* refers to functions that exist within a particular development project.)

3.4 The Entity-Relationship Data Model

The entity-relationship (E-R) data model is a widely used tool both in information analysis, systems analysis and database design. It is an outgrowth of the database design process and was developed initially in 1976. Its bases are the *entities* and *entity sets* discussed in the previous section, and *relationships* among the entity sets. Relationships, or more correctly relationship sets, are associations among entity sets. Throughout the remainder of this chapter the terms relationships and relationship sets will be used interchangeably, and will be distinguished by the use of italics. Entity sets will appear in caps.

As an illustration, CUSTOMERS, ORDERS and PRODUCTS may be three entity sets for an enterprise. Two obvious relationships among them may be expressed by the statement

CUSTOMERS *place* ORDERS *for* PRODUCTS.

This is shown in Figure 2 as an entity-relationship diagram.

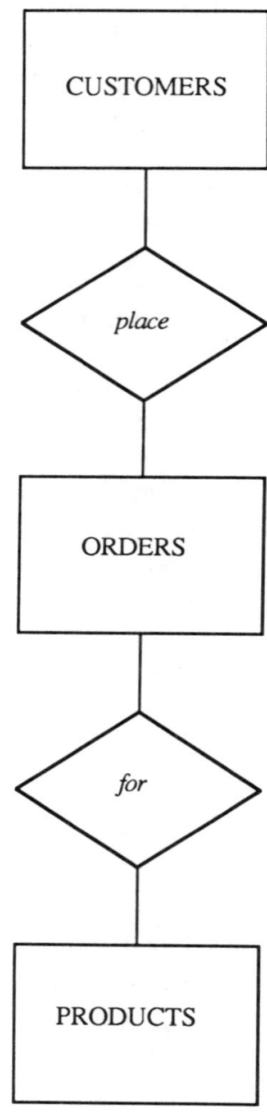

Figure 2: Entity-Relationship diagram: CUSTOMERS *place* ORDERS *for* PRODUCTS

Each entity set is denoted by a rectangle, and the diamonds depict relationships among the entity sets. Each entity set (CUSTOMERS, ORDERS, PRODUCTS) is the set of all customers, the set of all orders or the set of all products offered.

Within a university, some examples of entity sets may be STUDENTS, COURSES, FACULTIES and DEPARTMENTS. What are the relationships among them?

STUDENTS *enroll in* COURSES
FACULTIES *have* DEPARTMENTS
DEPARTMENTS *offer* COURSES

Notice that the entity sets are nouns and the relationships are verbs or prepositions, as shown in the E-R diagram in Figure 3.

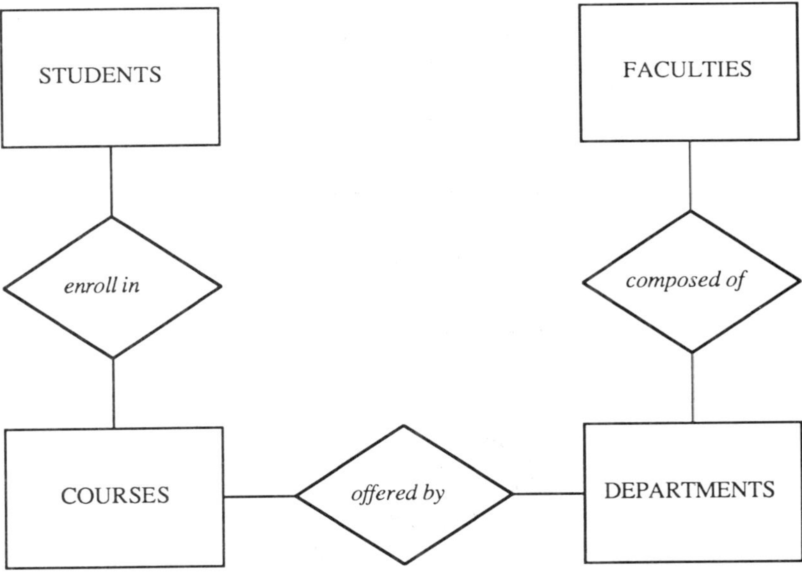

Figure 3: Partial E-R diagram of a university

The E-R diagram is also called the *conceptual schema*. *Conceptual* here means the highest level of abstraction – the way one ordinarily thinks about things. *Schema* means outline or diagram. It is useful to visualize data as having three levels of abstraction [9].

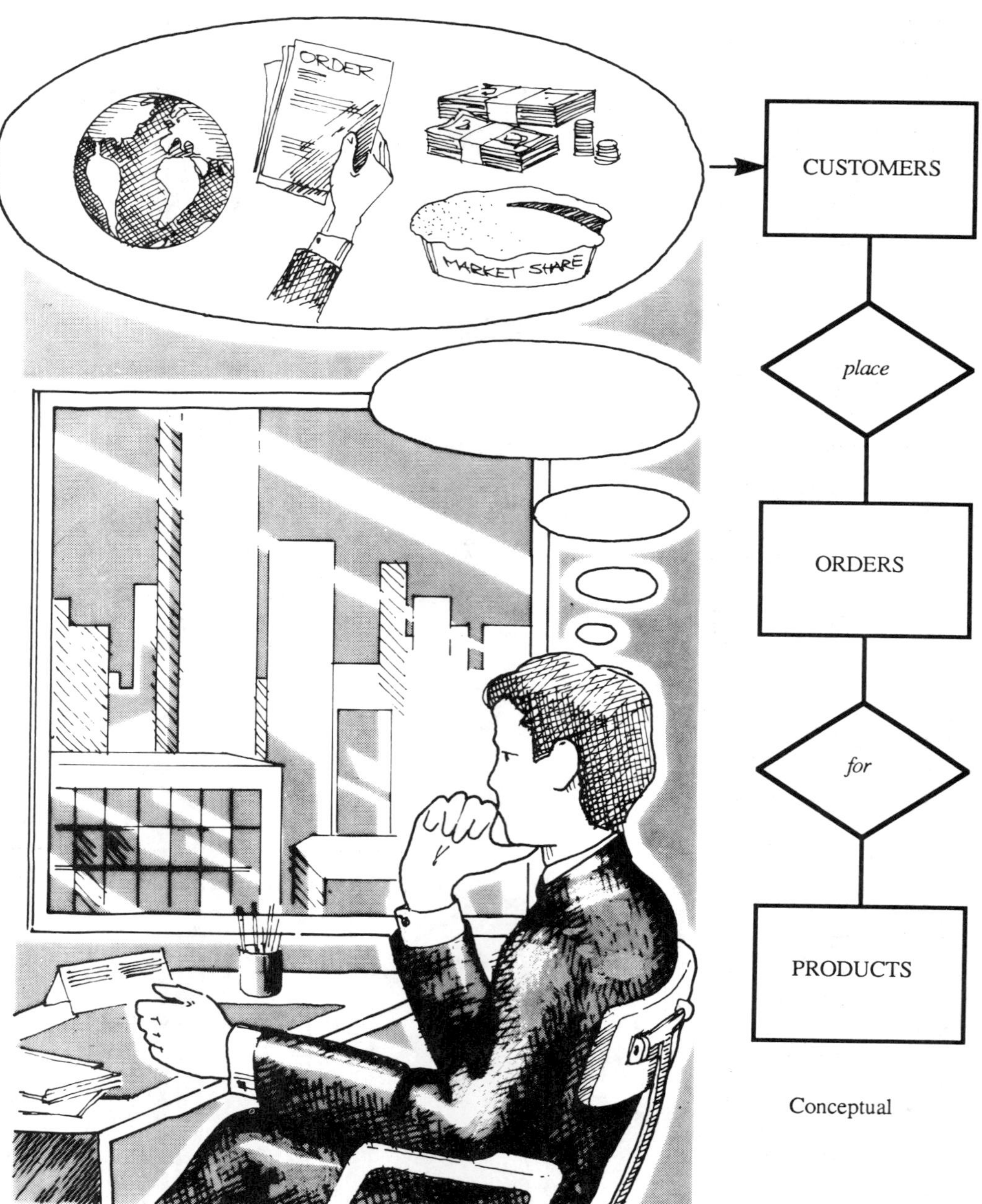

Figure 4: Three levels of data

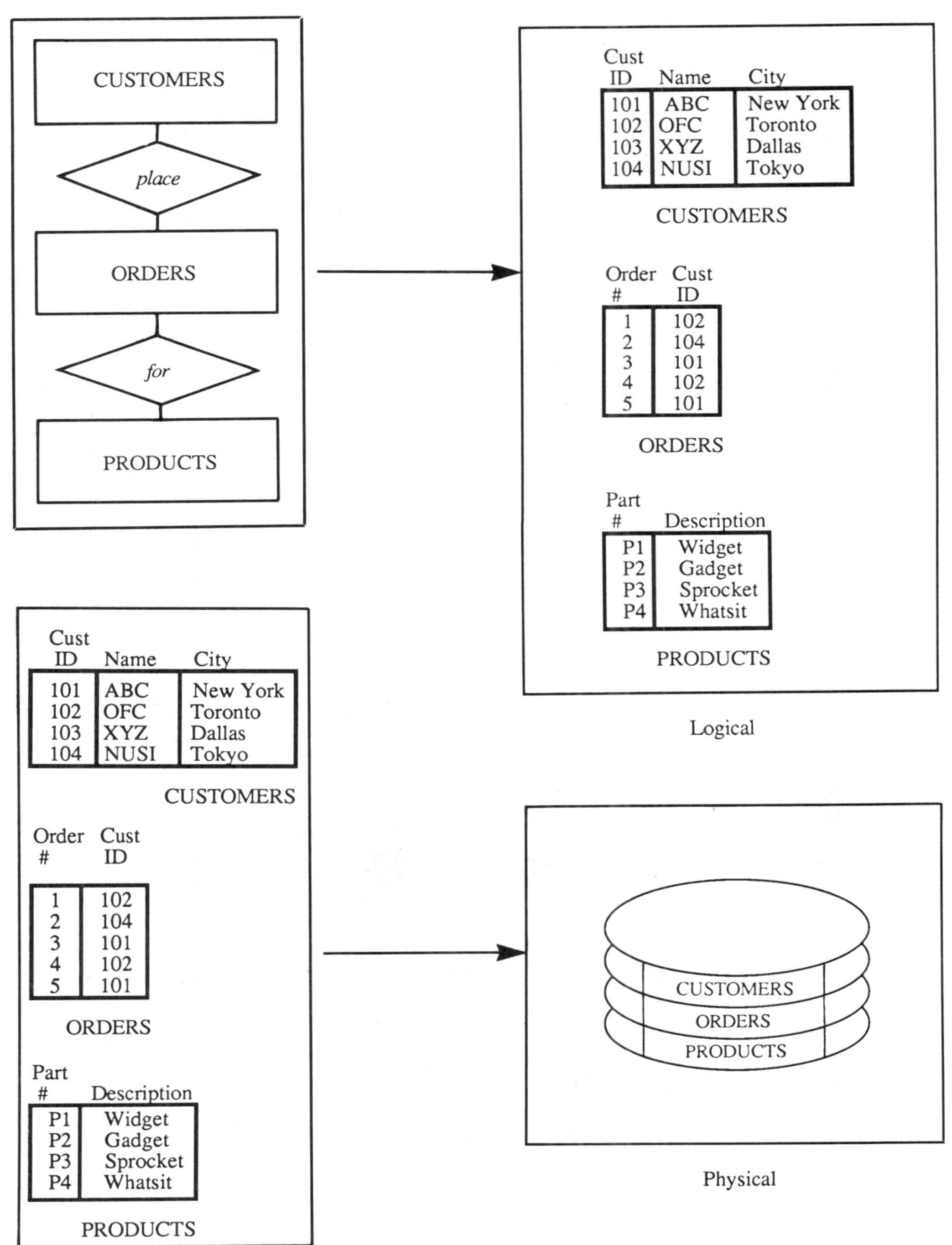
CUSTOMERS
place
ORDERS
for
PRODUCTS
Cust ID
Name
City
101 ABC New York
102 OFC Toronto
103 XYZ Dallas
104 NUSI Tokyo
CUSTOMERS
Order #
Cust ID
1 102
2 104
3 101
4 102
5 101
ORDERS
Part #
Description
P1 Widget
P2 Gadget
P3 Sprocket
P4 Whatsit
PRODUCTS
Logical
Cust ID
Name
City
101 ABC New York
102 OFC Toronto
103 XYZ Dallas
104 NUSI Tokyo
CUSTOMERS
Order #
Cust ID
1 102
2 104
3 101
4 102
5 101
ORDERS
Part #
Description
P1 Widget
P2 Gadget
P3 Sprocket
P4 Whatsit
PRODUCTS
CUSTOMERS
ORDERS
PRODUCTS
Physical

- The conceptual view is the highest level of abstraction, and corresponds to one's own perception of the data.

- The next lower level is the *logical level*, which is discussed in Chapter 4. In the logical view, one is concerned with the files, fields and so on. It provides the basis for database design.

- The lowest level, called the *physical* or internal level, is how data are actually stored in the computer. While this level is not of interest to an analyst, it may be of interest to database technicians and to systems programmers. (See Figure 4.)

(Although the treatment of data modeling in this text refers to a global data model, an E-R diagram has considerable use for individual software development projects. It is an excellent tool for analyst-user communication, and shows a high-level or *conceptual view* of data and data relationships.)

3.4.1 Relationships in the E-R Model

Of course, a student may enroll in several courses: English 101, Computer Science 225, History 438 and so on. As well, a particular course may have many students enrolled in it. For each student, one can associate all the courses in which he/she is enrolled, and for each course, one can associate all the students who are enrolled. We say there is a *many-to-many relationship* between STUDENTS and COURSES.

Each faculty has several departments: for example, within the faculty of Science, one may find the departments of Biology, Earth Sciences, Chemistry, Physics and Mathematics. Depending upon the organization of the university (called business rules or *constraints*), each department may (or may not) be within one faculty. Assume for now that each department belongs to one faculty. (Notice that discussion is limited to academic departments only, and administrative departments such as personnel, data processing, security, etc. have not been considered – a constraint of this particular E-R model.)

Each department offers several courses: English may offer English 101, 102, etc. However, a particular course is offered by only one department. These relationships (FACULTY to DEPARTMENT, and DEPARTMENT to COURSE) are described as being *one-to-many*. The nature of relationships can be added to an E-R diagram.

There are three types of relationships that may occur in E-R models:

- one-to-many (or many-to-one) relationships (1:N or N:1)

- many-to-many relationships (N:M or M:N)
- one-to-one relationships (1:1). (See Figure 5.)

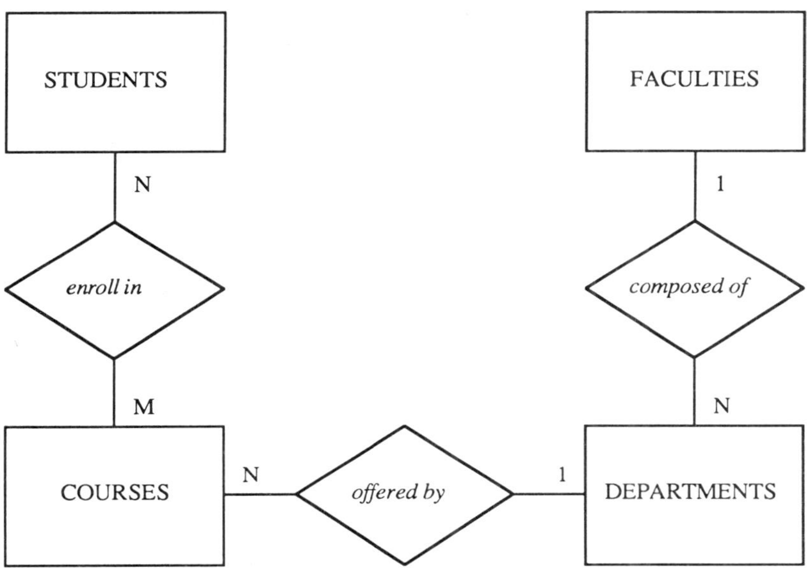

Figure 5: E-R diagram showing the nature of relationships

Marriage is an example of a one-to-one relationship occurring for the entity set PERSON (Figure 6).

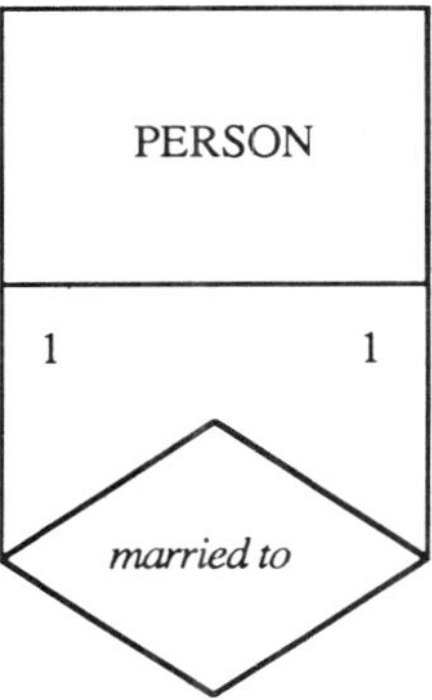

Figure 6: The relationship *marriage*

With respect to the university example discussed previously, consider the entity set PROFESSORS. Professors are those in a university who teach courses. For this example, assume that every department has a group of professors who work within that department only, and that among those professors, one is a department chairperson. Then the relationship associating each department with the chairperson of that department is a one-to-one relationship between

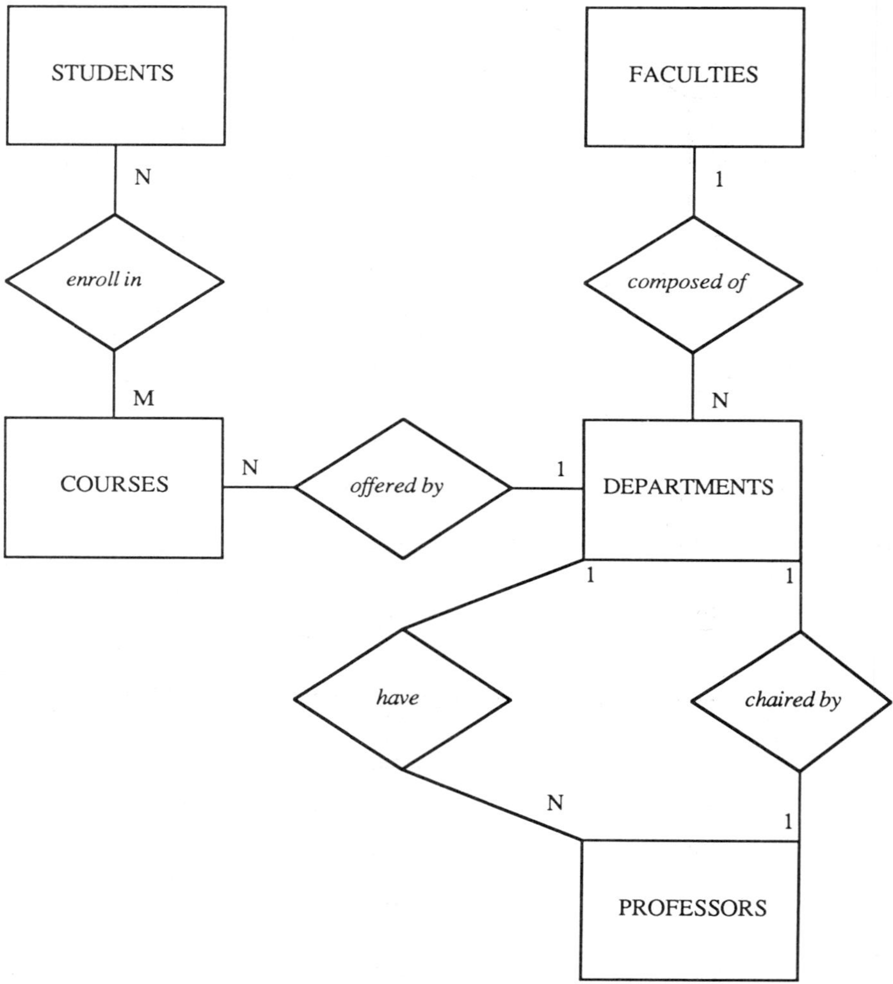

Figure 7: E-R diagram showing DEPARTMENTS and *chaired by*

DEPARTMENTS and **PROFESSORS**. These additions may be incorporated in the diagram. (See Figure 7.) Of course not all professors are department chair-people, so not all professors will participate in the *chaired-by* relationship. The professors who are department chairpersons are a *subset* of the set of professors.

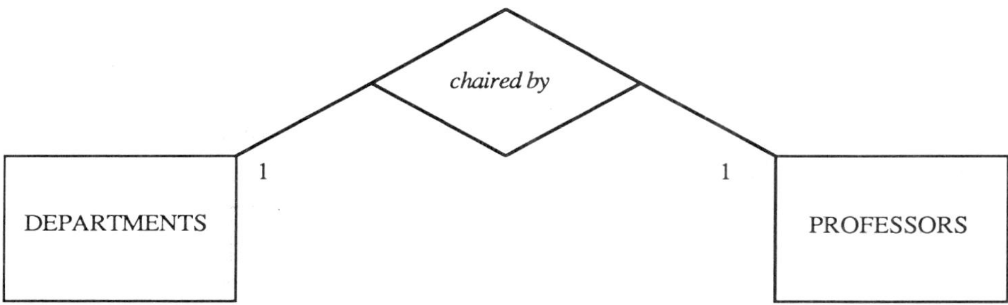

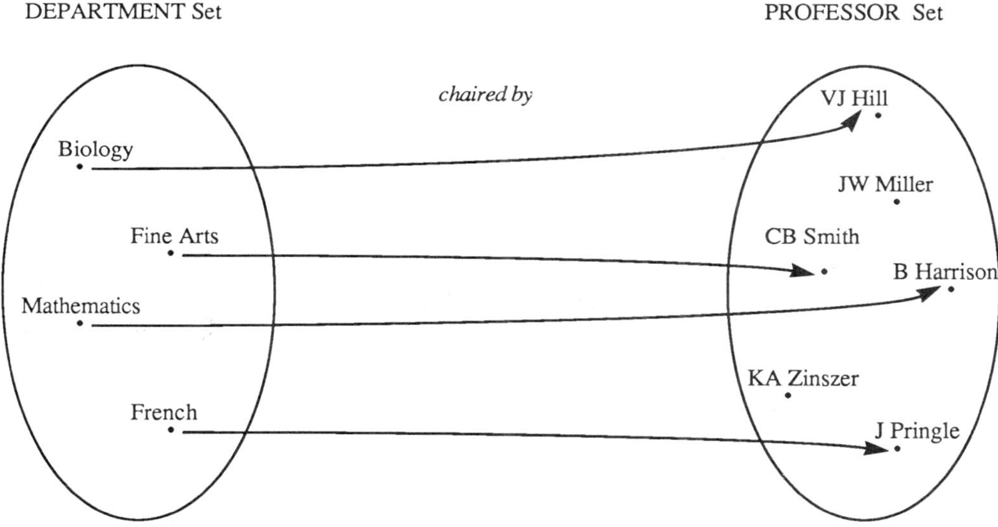

Figure 8: Chaired by

Some relationships are akin to the concept of mapping in mathematics. For example, consider the one-to-one relationship *chaired by* (Figure 8).

Each department is associated with the chairperson of that department. This can be written in mathematical notation as

chaired by: DEPARTMENT ----> PROFESSORS,

where *chaired by* is a mapping which associates with each department the chairperson of that department.

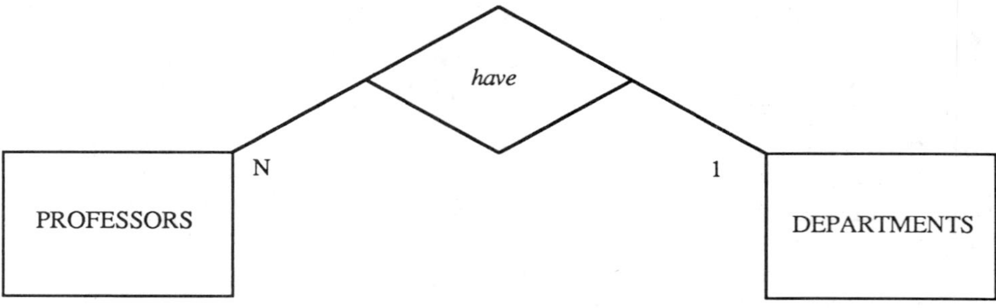

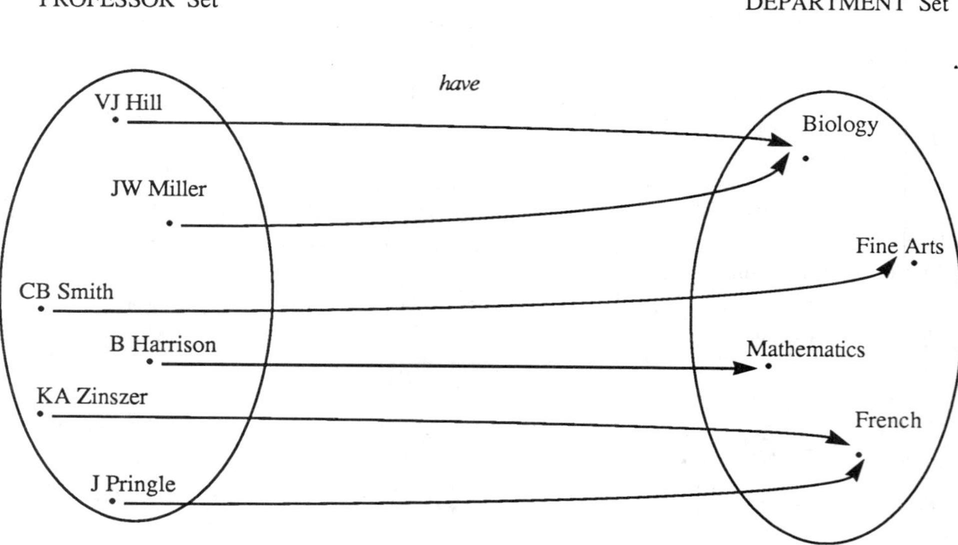

Figure 9: PROFESSORS *have* DEPARTMENTS

3.4.1.1 One-to-Many (or Many-to-One) Relationships

The one-to-many relationship *have*

PROFESSORS *have* DEPARTMENTS

is illustrated in the E-R diagram shown in Figure 9.

Have is a mapping which associates with each professor a particular department:

have: PROFESSORS ---> DEPARTMENT

This mapping is conceptually similar to that of a mathematical function. In mathematics, a function is used to describe a particular kind of association between two or more sets. More formally, a function f from set A to set B is a mapping which associates for every element of A, one and only one element of B. The authors have even borrowed the symbolism from mathematics, for it is common to write a function as

f: A ---> B.

The PROFESSOR set corresponds to A and the DEPARTMENT set corresponds to B. If the DEPARTMENT and PROFESSOR sets were interchanged, however, *have* would no longer be a function because for any department, *all* the professors in that department would be associated, as shown in Figure 10.

Relationships, then, are a generalization of the notion of a mathematical function. In fact, this very generalization frequently occurs in ordinary living. "The price of housing is a function of consumer demand," "a person's intelligence is a function of heredity" and "the length of the ski season is a function of the weather" are three such illustrations. In each illustration, the use of the word function implies an association between two changing events or variables, one of which determines the other (the other being dependent upon the one). One event is an *independent variable* and the other is *dependent*. The notion of dependency becomes important when considering data at the logical level (Chapter 4) because some data dependencies behave poorly under automation.

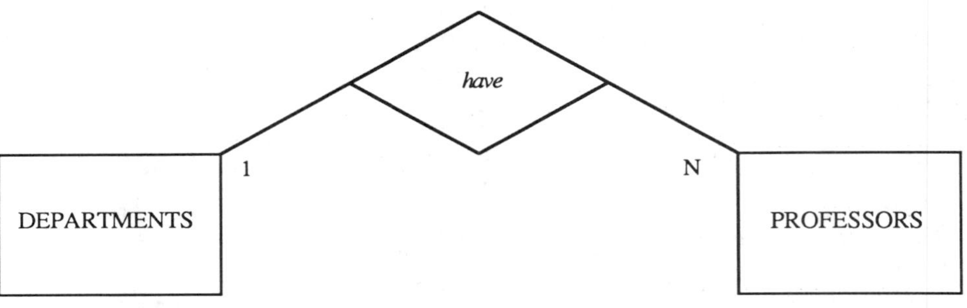

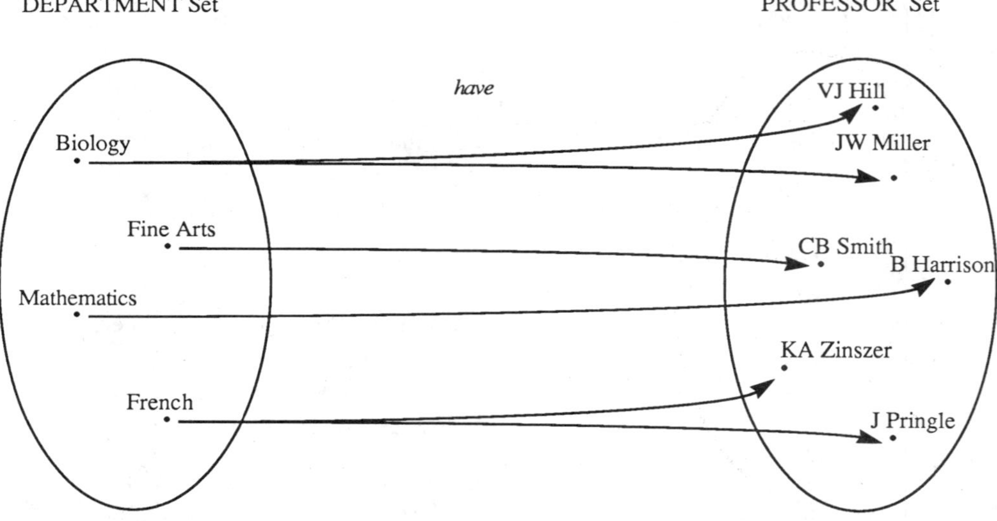

Figure 10: DEPARTMENTS *have* PROFESSORS

3.4.1.2 Many-to-Many Relationships

The most complex of the three relationships is many-to-many. An entity in an entity set A may be associated with any number of entities in entity set B, and conversely an entity in B may be associated with any number of entities in A.

STUDENTS *enroll in* COURSES

is an example of a many-to-many relationship. (See Figure 11.)

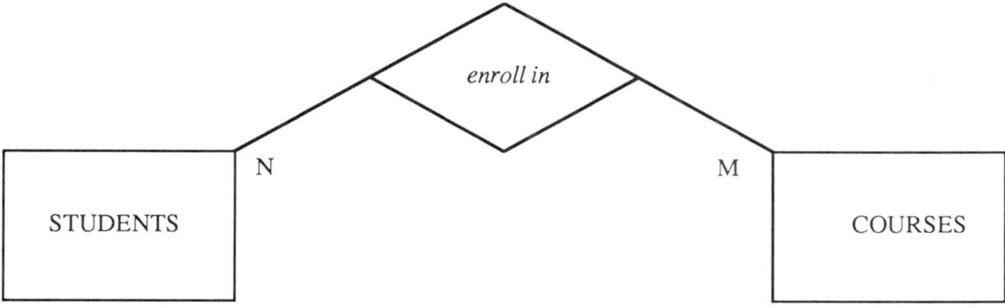

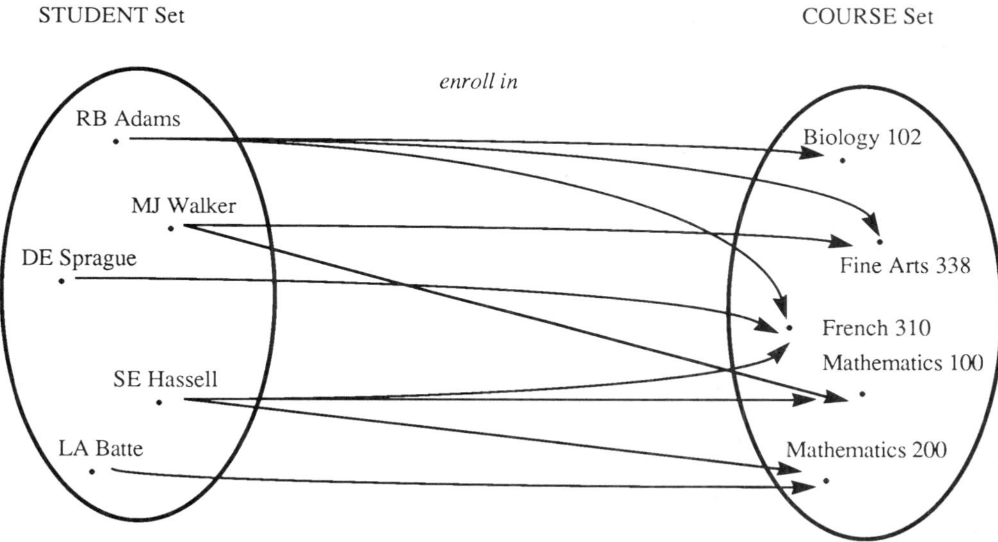

Figure 11: STUDENTS *enroll in* COURSES

3.4.2 Attributes of Entities and Relationships

Entity sets and relationship sets have perceived properties called *attributes*. The attributes of a student may be name, student number, address and major program of study. In fact, when describing entities and relationships one usually describes them in terms of such properties. Attributes of the entity COURSES may include course number, course description and credit hours.

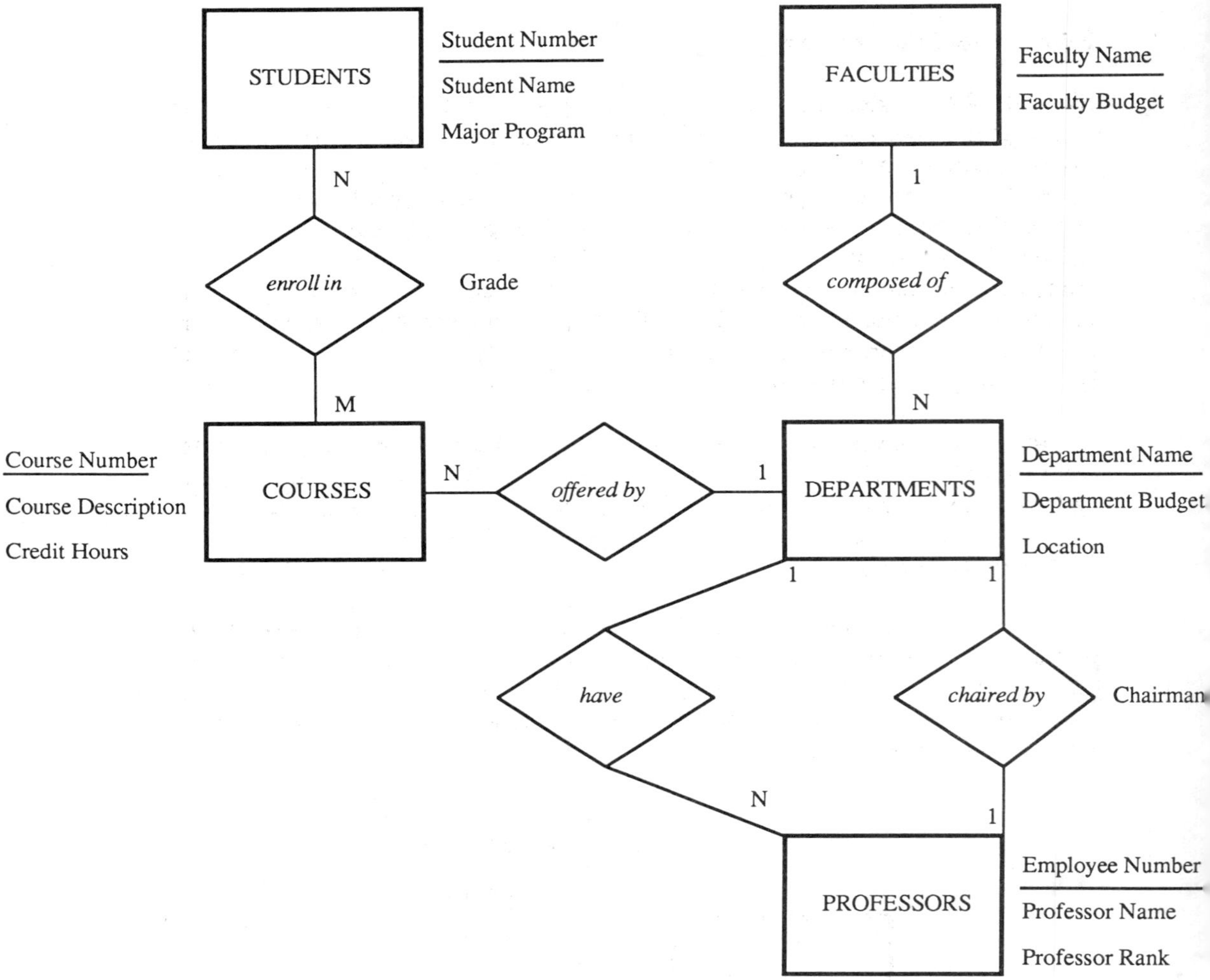

STUDENTS
Student Number
Student Name
Major Program
N
enroll in
Grade
M
FACULTIES
Faculty Name
Faculty Budget
1
composed of
N
Course Number
Course Description
Credit Hours
COURSES
N
offered by
1
DEPARTMENTS
Department Name
Department Budget
Location
1
1
have
chaired by
Chairman
N
1
PROFESSORS
Employee Number
Professor Name
Professor Rank

Figure 12: Attributes of STUDENTS and COURSES

The relationship set *enroll in* associates students with the particular course they are taking. *Enroll in* has the attribute Grade, which means the grade received by a particular student for a particular course. In general, relationship sets may or may not have attributes (apart from the attributes implied by their connections with associated entities – in this case Student Number and Course Number).

Attributes may be added to an E-R diagram, as shown in Figure 12.

3.4.3 Identifier Attributes

The E-R data model attempts to represent relevant data in a way that is both meaningful to users and conducive to automation. To this end, a way is needed to distinguish entities and relationships from one another in terms of their attributes. This is done through the use of attributes that can identify each entity or relationship. Student Number is an example of such an attribute for the entity set STUDENT. Each student is distinguishable from all other students by a unique student number. Course Number (e.g., English 102) is an example of an identifying attribute for the entity COURSES. Employee Number, Part Number, Order Number, Invoice Number, Social Security Number, are examples of identifying attributes from various applications areas.

In a large university, Student Name may not be an appropriate choice for the identifying attribute of the entity STUDENT, because two students may have the same name. The concept of identifying attributes is analogous to that of *primary key* in data processing, and in discussions the two terms will be used interchangeably.

It is often the case that there is more than one possible *candidate* for the primary key. As an illustration, consider an entity set EMPLOYEES, which concerns the employees of a company. Attributes may include Employee Number, Employee Name, Employee Address, Title, Salary and Social Security Number, as shown in Figure 13.

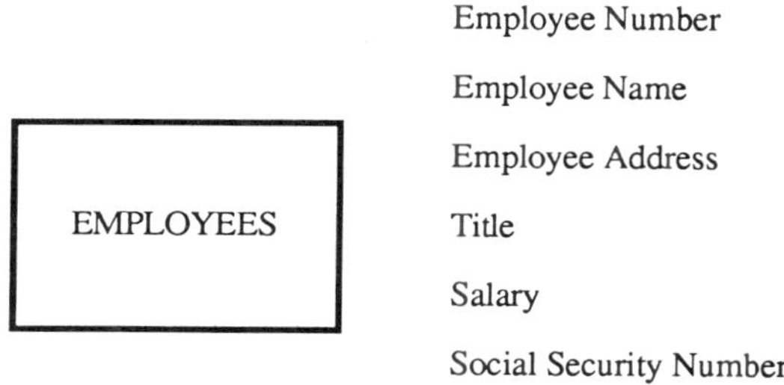

Figure 13: EMPLOYEE with more than one candidate key

If one makes the assumption (constraint) that no two employees who have the same name live at the same address, then there are three candidates possible for the primary key of EMPLOYEES:

Employee Number

Social Security Number

Employee Name plus Employee Address.

If such a situation exists, one of the *candidate keys* must be selected as the primary key. A common convention is to underline primary keys.

For relationships, the primary key is obtained by choosing the identifying attributes of the participating entities. For example, the primary key of *enroll in* is

Student Number plus Course Number.

Because this is always the case, the identifying attributes of relationships are often omitted in an E-R diagram.

3.4.4 Changing Relationships into Entities

One weakness with the E-R model is that there is no way of expressing relationships among relationships, a relationship being defined as an association among entities. Assume that in the university example described in this chapter, courses are offered during three semesters: fall, winter and spring, and that SEMESTERS should be an entity within the model. Perhaps not all courses are offered every semester. As the model stands in Figure 12, there is no "legal" way of relating professors to the courses *offered by* departments, because no relationship can exist between a relationship (*offered by*) and an entity (PROFESSORS). However, both entity sets and relationship sets are describable by attributes, and because they are conceptually rather similar, why not circumvent this limitation by changing relationships into entities! In fact, this is often done in practice.

Figure 14 shows the addition of the entity SEMESTERS to the model as well as the relationships

enroll in

offered by

chaired by

changed into entities.

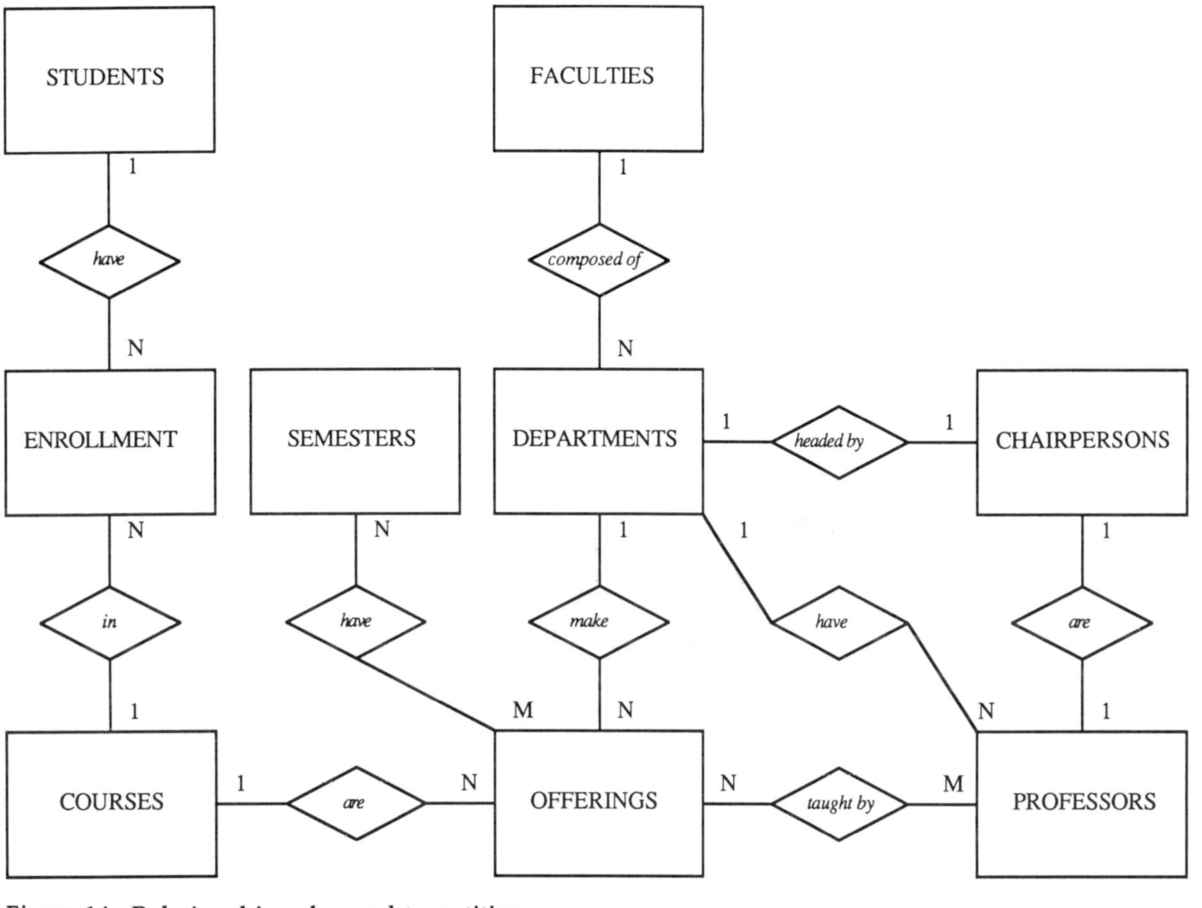

Figure 14: Relationships changed to entities

Notice that by changing the many-to-many relationship *enroll in* into the entity ENROLLMENT, it is decomposed into two simpler relationships:

- a one-to-many relationship (STUDENTS *have* ENROLLMENT)
- a many-to-one relationship (ENROLLMENT *in* COURSES).

One can generalize, and state that a many-to-many relationship can always be decomposed into two relationships (a one-to-many relationship and a many-to-one relationship). Because one-to-many relationships are conceptually easier, such decomposition is in fact desirable.

The relationship *offered by* has been changed to the entity OFFERINGS in order to express relationships between SEMESTERS and OFFERINGS, and between OFFERINGS and PROFESSORS. However, the resulting relationships are not simpler: they are many-to-many. It is left as an exercise to further refine the model so that these relationships are each decomposed into two one-to-many relationships.

3.4.5 Additional Characteristics of Entities and Relationships

There are additional characteristics of entity sets and relationship sets that are useful to consider in information systems, and that can be accommodated with an E-R model. In this section, three such characteristics are explored.

3.4.5.1 Existence Dependency: Strong and Weak Entities

Consider again the E-R diagram shown in Figure 2, which depicts the entity sets CUSTOMERS, ORDERS, and PRODUCTS, and the relationships

CUSTOMERS *place* ORDERS *for* PRODUCTS.

The entity set PRODUCTS is called a *strong* entity set because it exists independently of the entity sets CUSTOMERS and ORDERS. A company may offer a product that has never (to date) been sold to a customer, a product for which there has never been an order associated. (Of course, one expects that eventually an order will exist for such a product; otherwise the product would be discontinued.) However, an order cannot exist unless it is for a product offered by the company; its existence depends on the entity set

PRODUCTS. Its existence is also dependent upon the entity set CUSTOMERS, for an order must be from a customer. Assume for discussion purposes that the entity set CUSTOMERS is meant to include those potential customers who as yet may not have actually purchased products – those from whom future orders are anticipated. Then within this framework, CUSTOMERS and PRODUCTS are strong entity sets. ORDERS is a weak entity set, and owes its existence to both CUSTOMERS and PRODUCTS. Weak or existence-dependent entities are indicated by a double rectangle and by an arrow pointing toward the dependent entity. (See Figure 15.)

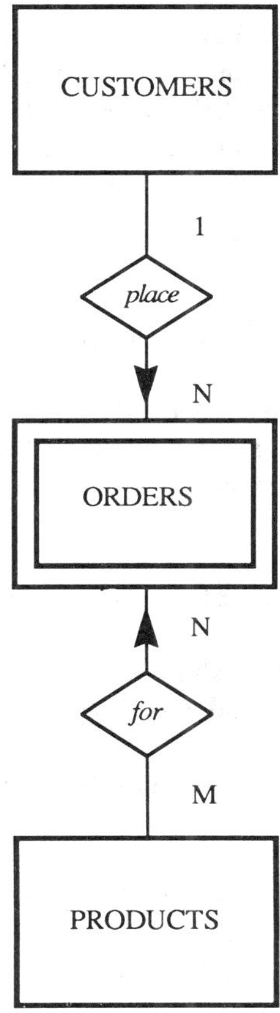

Figure 15: Strong and weak entity sets

3.4.5.2 Optional and Mandatory Participation in Relationships

There may be some (potential) customers who have never placed an order, and there may be some products that have never been sold. For such customers and such products, there is no participation in the relationships

CUSTOMERS *place* ORDERS *for* PRODUCTS.

This refinement of optional participation in a relationship can be shown in an E-R diagram by circling the line joining the "optional" entity set and the relationship (Figure 16).

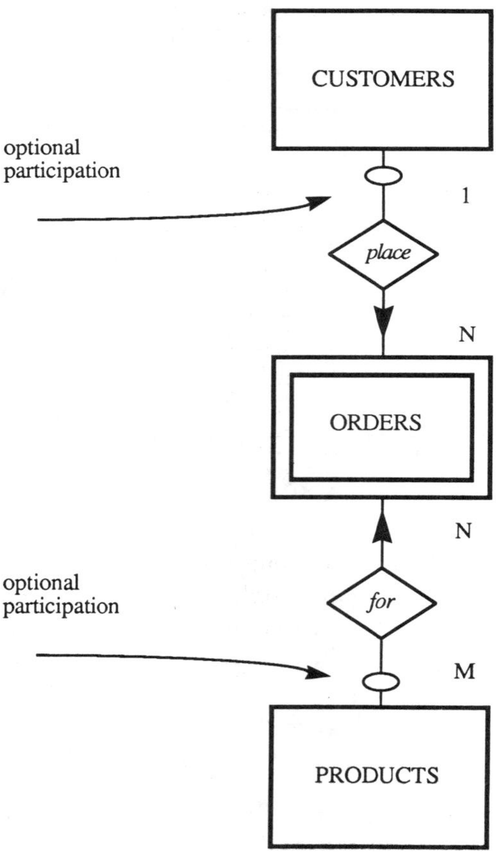

Figure 16: Optional and mandatory participation

If participation in a relationship is not optional, then it must be mandatory. Mandatory participation is left unlabeled, and is implied by the absence of the optional indicator.

Returning to the earlier example of the university E-R model, not all professors are department chairpersons, so participation of the entity set PROFESSORS in the relationship

PROFESSORS *are* CHAIRPERSONS (Figure 14)

can be characterized as optional.

(Notice that the E-R model cannot completely capture the world it attempts to model, for although participation of PROFESSORS in the relationship

PROFESSORS *are* CHAIRPERSONS

is optional, one presumes that every department must or does have a chairperson, so there are some professors for whom participation in the relationship is mandatory. Although E-R models cannot capture all of reality, they represent a quantum leap relative to earlier modeling tools.)

The optional or mandatory nature of participation in a relationship depends upon the particular business policies or business rules of an enterprise – the *constraints* of a particular E-R model. In the previous example of CUSTOMERS, ORDERS and PRODUCTS, it was stated that products may exist that have not yet been sold. To state such a policy presumes knowledge of the business area under investigation.

If one's business is to custom-produce products for customers, then it may be that no product exists unless it has been ordered by a customer. If this were the case, not only would the participatory nature of PRODUCTS in the relationship

ORDERS *for* PRODUCTS

change from optional to mandatory, but also the existence dependency of ORDERS on PRODUCTS would change to PRODUCTS on ORDERS (Figure 17). (This permits the question "Can an order exist without a product?" and opens the door on semantics (meaning of data) and semantic modeling generally, which is beyond the scope of this text.)

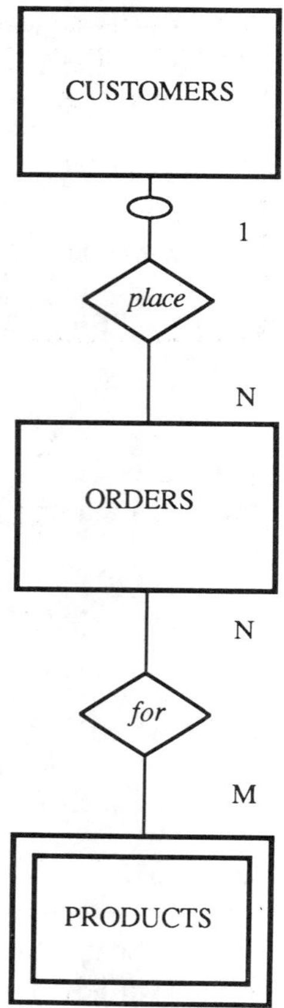

Figure 17: Existence dependency of PRODUCTS on ORDERS

3.4.5.3 Generalization and Specialization: Subtypes and Supertypes

The entity set CUSTOMERS described in the example of the previous section is an amalgamation (or *union* in set theory) of two kinds of customers:

- customers who have not yet ordered products (potential customers)
- customers who have at some time ordered products (current customers).

Then the entity set CUSTOMERS is a *generalization* of the entity sets POTENTIAL CUSTOMERS and CURRENT CUSTOMERS. Generalization can be accommodated with E-R diagrams (Figure 18).

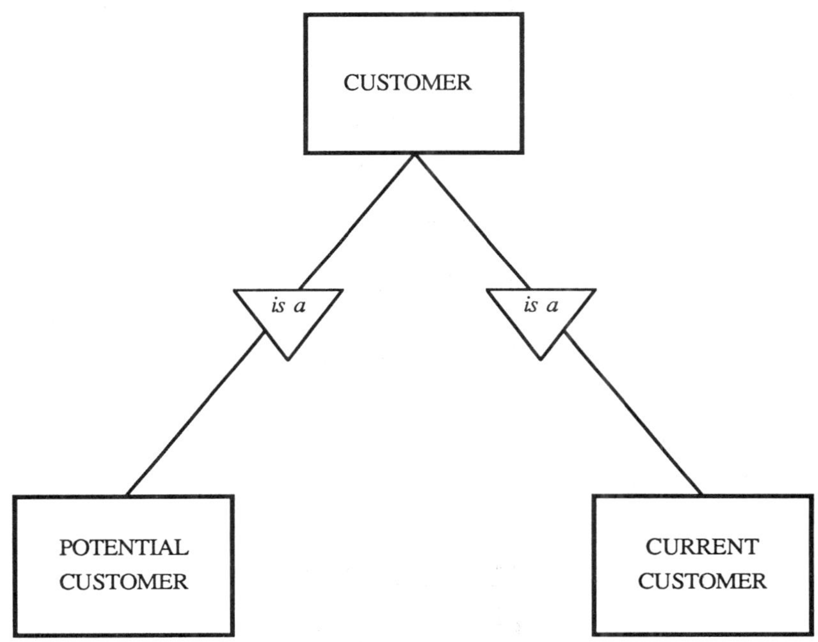

Figure 18: Generalization and specialization

A potential customer or a current customer is called a *subtype* of CUSTOMER, and conversely a customer is a *supertype* of POTENTIAL CUSTOMER and of CURRENT CUSTOMER. Both POTENTIAL CUSTOMERS and CURRENT CUSTOMERS represent specialization of the entity set CUSTOMERS.

3.4.6 Data Modeling: Semantics, Syntax and Constraints

E-R diagrams can provide a global data model. However, for any particular software project, they can also be a developer-user communications tool. In addition, because E-R modeling developed "from the bottom up," it can be translated readily into a form amenable to automation. As one translates to the lower level, the constraints of the model become more mechanical, and provide the rules for manipulating the data.

For the sake of clarity, we define three commonly used data-modeling terms. The term *semantics* refers to the meanings attached to terms used in the model, the term *syntax* refers to compliance to the rules of the general data model being used, while the term *constraints* refers to the particular business rules or policies of the area under investigation.

For example, one syntactical consideration of the E-R model is that relationships are associations among entities, and are therefore *not* permitted among relationships. Then, to conform to the syntax of the E-R model, one must abide by this rule (and change some relationships to entities). Some constraints particular to any specific E-R model correspond to the appropriate business rules and policies of the enterprise. Examples have been developed that illustrate this:

- for the university example,
  - departments are located within faculties,
  - students are those who are currently enrolled in at least one course
- for the business example,
  - customers include both potential and current customers.

In the process of translating to a lower level of abstraction, the meanings or semantics become separated from the syntax, and become less important. What matters is syntax: not violating the model, and imposing a rigid, mechanized structure that can be accommodated by automation. The meaning or interpretation rests then with the user.

3.5 Exercises

1. Refine the E-R diagram of Figure 14 to include specialization of professors: adjunct, assistant, associate and full professors. As well, make the

changes necessary for changing the many-to-many relationships into one-to-many relationships. How effective is the E-R model in handling these aspects?

2. Suppose the business rules associated with the scenario for the E-R diagram shown in Figure 2 include the following:

   - Each product is supplied by one supplier, but a supplier may supply many products.
   - Products can either be components that require assembly into finished products by the company's employees, or finished products ready to sell.
   - Employees may have special skills used for assembling components into final products.
   - Salespeople are assigned territories, and call on customers within their particular territories. Salespeople are chosen for their expertise in various areas (microelectronics, hydraulics, hydrogeology, etc.) and each salesperson is restricted to a particular product line reflecting that expertise. The result is that different salespeople may call on the same customers, but to sell different products.

   Expand and refine the E-R diagram of Figure 2 to include the above considerations.

3. One of the more significant information systems projects developed at the University of Waterloo was the automation of the Oxford English Dictionary. In May of 1984, the Oxford University Press formally announced its intention to computerize the Oxford English Dictionary (OED). In late 1987, after over 200 person years of effort, much of which was devoted to data entry of more than 60 million words, the Oxford University Press announced that the OED and its supplement were available in optical disk format. A database design was developed by the University of Waterloo for this integrated version of the OED. Since the announcement, there has been much enthusiasm and interest in this project throughout North America, Europe, and the Far East.

   To illustrate the difficulty in designing a lexical database, consider the following definition from the OED:

Dictionary.
1. A book dealing with the individual words of a language (or certain specified classes of them), so as to set forth their orthography, pronunciation, signification, and use, their synonyms, derivation, and history, or at least some of these facts: for convenience of reference, the words are arranged in some order, now, in most languages, alphabetical; and in larger dictionaries the information given is illustrated with quotations from literature; a word-book, vocabulary, or lexicon.

2. fig. A person or thing regarded as a repository of knowledge, convenient for consultation.

Dictionaries have typically been considered to be alphabetically arranged books. The electronic form of dictionary, on the other hand, is viewed as a "repository of knowledge, convenient for consultation" and attempts to capture the *deep structure* of the knowledge base, often only implicitly present in the printed version, without losing the explicit form of expression, meticulously crafted by its lexicographers. The new OED is also *simultaneously* capable of producing galley proofs for the traditional book form in addition to being available in electronic form. It is believed that much etymological knowledge will be gained by having the new OED available in such a convenient format.

Derive a conceptual model of a dictionary (lexical database), and document it using an E-R diagram.

4. Fly-High Airlines provides flights to a number of small towns in the Midwest and to the main connecting ports for the major airlines. The company has decided that in order to remain competitive, it must redesign its passenger reservation system to provide effective support for flight reservations by passengers for both Fly-High Airlines and connecting airlines. Information needs in the following five areas have been identified as being important:

- Reservation requests and inquiries by customers
- Flight inquiries by customers regarding schedules to and from destinations served by Fly-High Airlines and other airlines
- Flight scheduling, creation, and deletion by airline flight administrators
- Plane allocation or reallocation to flights by airline plane administrators
- Seat allocation to reservations by check-in staff.

Many rules govern how a business operates. Some of them originate with government regulations, some simply reflect the natural laws that govern people and their machines, and others are mandated by senior management who really know how the business should be run. A few rules have their origins in history, and no one quite knows why they exist. The following rules must be enforced by the system.

- Reservations requested must be immediately accepted, rejected or wait-listed.
- A group of passengers may make a group reservation, which is then accepted or rejected as a whole, according to whether sufficient unused seating capacity remains. Although a group reservation is accepted or rejected as a whole, it is made up of single reservations, each of which must have a passenger identified with it.
- A passenger may not hold conflicting reservations, such as being on different flights at the same time, or holding more than one reservation on the same flight.
- Flight departure time from the departure airport must be before flight arrival time at the arrival airport, and flying time must satisfy estimates.
- Planes cannot be allocated to flights with conflicting schedules.
- A flight has no intermediate airports.
- Each flight has a nominal seat capacity and a nominal seat margin associated with it. This means that last minute cancellations and no-shows are anticipated; the margin is calculated as a percentage of the nominal seat capacity. The capacity of the plane first allocated must exceed or equal the nominal seat capacity.
- The number of reservations for a flight cannot exceed the nominal capacity plus margin. The nominal capacity becomes the actual capacity when a plane is allocated.
- At certain times before a flight, the total reservations held for a flight are examined by plane allocators. If defined maximum thresholds have been reached, reallocation to a larger plane can be triggered. Similarly, if defined minimum thresholds have been reached,

reallocation to a smaller plane may be made. Plane allocators, who must combine the knowledge of airport controllers and fuel suppliers with the cunning of successful hot dog vendors at baseball games, are paid according to revenue gained by *optimally* matching planes to reservation demand. Roughly speaking, optimal means the best that has been achieved in the past.

- Seat allocation for reservations is not completed until a certain time (generally 15 minutes) before flight departure.

a. Analyze the scenario described above and document it using an E-R diagram.

b. Recall that the functions of an information system are the dual of the data component, and are meaningful only if the proper data are present. Now that you have modeled the data for Fly-High airlines, can you think of business functions from the five areas mentioned previously that the E-R model can support? One example might be "Group reservation acceptance, rejection, or wait-listing."

5. A public library wants to keep track of books, patrons, loans, etc. The following is a description of some business rules about the library:

- Books have a code, called an access code, which indicates the location in the library shelves. As well, each book has an inventory number, assigned when a book is purchased. Every book, of course, has a title, one or more authors, a publisher and date of publication. For some books, the library has more than one copy. In such cases, all copies have an access code that differs only by a suffix "c.1," "c.2," etc.; however, the inventory numbers for such books may be unrelated. Most books are allowed to circulate, but some of them (e.g., encyclopedias) can be used only in the library's reading room.

- Users of the library have a number, name, address and telephone.

- Users with overdue books must pay fines. A user must pay all outstanding fines before being allowed to borrow more books.

- Users who find that a book is unavailable can ask to be put on a waiting list for that book. The list is implemented on a first-come first-serve basis. When the book becomes available, the first user is notified, and the book is held for five days. If it is not picked up during this period of time, the next user is notified, and so on.

a. Develop a conceptual data model for the library, and document it using an E-R diagram. For each entity type and relationship type, include a brief statement indicating the role played in the system. State any additional assumptions you feel are necessary.

b. After asking you to work on this project, the library decides to start a record collection. In many ways, records are similar to books. As well, to evaluate the use of its collection, the library wishes to know the number of times each record has been borrowed. Discuss how the addition of this new material affects your model, and redraw your E-R diagram to accommodate such changes.

6. Explain the difference between generalization and specialization. Illustrate using examples.

7. Why are semantics more important when considering higher levels of abstraction? Provide examples to support your argument.

8. The Campus Bookstore is trying to establish a system to co-ordinate the ordering of books used by professors in their courses. Develop an E-R model that could represent this situation. State the business rules you require.

9. A company builds products to order for its customers. Certain employees, depending upon their skills, will work on particular projects, and each project has a manager who is responsible for overseeing the project. Parts for each project are purchased from suppliers, and one part may be purchased from several different suppliers. Develop an E-R model to represent such an enterprise.

10. The Metropolitan Transportation Authority wants to improve its information services to customers, drivers and managers. Buses need to be better scheduled for repairs, as many unionized mechanics now spend much of their working day reading magazines because all buses are in service. Buses operate around the clock. Currently, drivers can ascertain which shifts they will be working by looking at a roster posted in the station. Timetables showing bus routes and times of arrival and departure at each bus stop are available with each bus driver. As well, the timetables show the

fares that apply. However, many prospective passengers telephone the Transportation Authority directly to obtain information, rather than going to the trouble of obtaining a timetable.

Develop an E-R diagram that defines your concept of a bus information system. State the business rules for your system.

11. General Hospital is a medium-sized hospital in the center of the city. It wishes to improve its information system regarding work scheduling for nurses, bed allocation and future projections of bed availability. Each department of the hospital (physiotherapy, recovery, intensive care, etc.) has nurses assigned to it. Nurses may work any or all shifts. Each ward in the hospital has patients, and each patient may be assigned several nurses.

    Prepare an E-R diagram that corresponds to your idea of the scenario described above.

12. How effective is the E-R model in capturing the dynamics of a system over time? What generally are the positive aspects of the E-R model? What are its limitations? Support your discussion with examples.

3.6 Bibliography

1. Albano, A.V. de Antonellis, and di Leva, A. *Computer-Aided Database Design* Amsterdam, The Netherlands. North-Holland, 1985.

2. Armitage, Howard M. and McCarthy, William E. *Decision Support Using Entity-Relationship Modeling. Journal of Accounting and EDP.* Fall, 1987.

3. Biggs, Norman L. *Discrete Mathematics.* London, England. Clarendon Press, 1985.

4. Bradley, James. *An Introduction to Data Base Management in Business, Second Edition.* New York, New York. Holt, Rinehart and Winston, 1987.

5. Brodie, M.L., Mylopoulos, J. and Schmidt, J.W. *On Conceptual Modeling: Perspectives from Artificial Intelligence, Databases, and Programming Languages.* New York, New York. Springer-Verlag, 1984.

6. Chen, Peter P. (editor). *Entity-Relationship Approach to Information Modeling and Analysis*. Amsterdam, The Netherlands. North-Holland, 1983.

7. Chen, Peter. *The Entity-Relationship Approach to Logical Data Base Design*. Q.E.D. Monograph Series, No. 6. Wellesley, Massachusetts. Q.E.D. Information Sciences, 1977.

8. Chen, P.P.S. *The Entity-Relationship Model - Towards a Unified View of Data. ACM TODS 1 No. 1.*, 1976.

9. Dampney, C.N.G. et al. *Data (Information) Analysis of Business Systems - Informal Notes*. Waterloo, Ontario. University of Waterloo, 1987.

10. Date, C. J. *An Introduction to Database Systems, Volume 1, Fourth Edition*. Reading, Massachusetts. Addison-Wesley, 1986.

11. Davis, G.B. *Management Information Systems - Conceptual Foundations, Structure and Development*. New York, New York. McGraw-Hill, 1983.

12. Furtado, A.L. and Neuhold, E.J. *Formal Techniques for Data Base Design*. Berlin, West Germany. Springer-Verlag, 1986.

13. Gray, Peter. *Logic, Algebra and Databases*. Chichester, U.K. Ellis Horwood, 1984.

14. Griethuysen, J.J. van (editor). *Concepts and Terminology for the Conceptual Schema and the Information Base*. ISO TC97/SC5/WG3, New York, New York. International Organization for Standardization, 1982.

15. Howe, D.R. *Data Analysis for Data Base Design*. London, U.K. Edward Arnold, 1983.

16. Jackson, M.A. *System Development*. Englewood Cliffs, New Jersey. Prentice-Hall, 1983.

17. Korth, Henry F. and Silberschatz, Abraham. *Database System Concepts*. New York, New York. McGraw-Hill, 1986.

18. Langefors, B. *Information Systems Theory*. Information Systems, Vol. 2, pp. 207-219, 1977.

19. Martin, James. *Managing the Data Base Environment*. Englewood Cliffs, New Jersey. Prentice-Hall, 1983.

20. Norris, Fletcher R. *Discrete Structures: an introduction to mathematics for computer science*. Englewood Cliffs, New Jersey. Prentice-Hall, 1985.

21. Parkin, A. *Systems Analysis*. London, U.K. Edward Arnold, 1980.

22. Parkinson, Richard C. *Data Analysis: the Key to Data Base Design*. Wellesley, Massachussetts. QED Information Sciences, 1985.

23. Sowa, J.F. *Conceptual Structures: Information Processing in Mind and Machine*. Reading, Massachusetts. Addison-Wesley, 1984.

24. Tsichritzis, Dionysios and Lochovsky, Frederick H. *Data Models*. Englewood Cliffs, New Jersey. Prentice-Hall, 1982.

Chapter 4

THE RELATIONAL MODEL

4.1 Introduction

The previous chapter dealt with the topic of corporate data modeling using the entity-relationship approach. This chapter concerns the translation of the E-R model to a form amenable to automation, and the refinements necessary to the conceptual model for the translation. The objective of this chapter is to provide some basic knowledge of databases, which form a major component of many, if not most, information systems. In particular, the relational database model is discussed.

4.2 Objectives of the Automated Data Model

The entity sets of the E-R model discussed in the previous chapter are those things about which information is desirable. For a company, these things may be customers, orders and products. For a university, students, courses, departments and faculties are probably important. These entities will eventually take the form of storage files on a computer. After the data for the enterprise has been conceptually modeled, the model can be converted to the logical level, called the relational data model. (See Figure 19.)

The goal in translating entity sets to a form more conducive to automation is to produce a set of logical files (a database) that consist of exactly what is needed to satisfy users and that behave well under automation. To this end, one strives for the simplest possible system, for one in which redundancy is controlled and for which there are few (if any) anomalies.

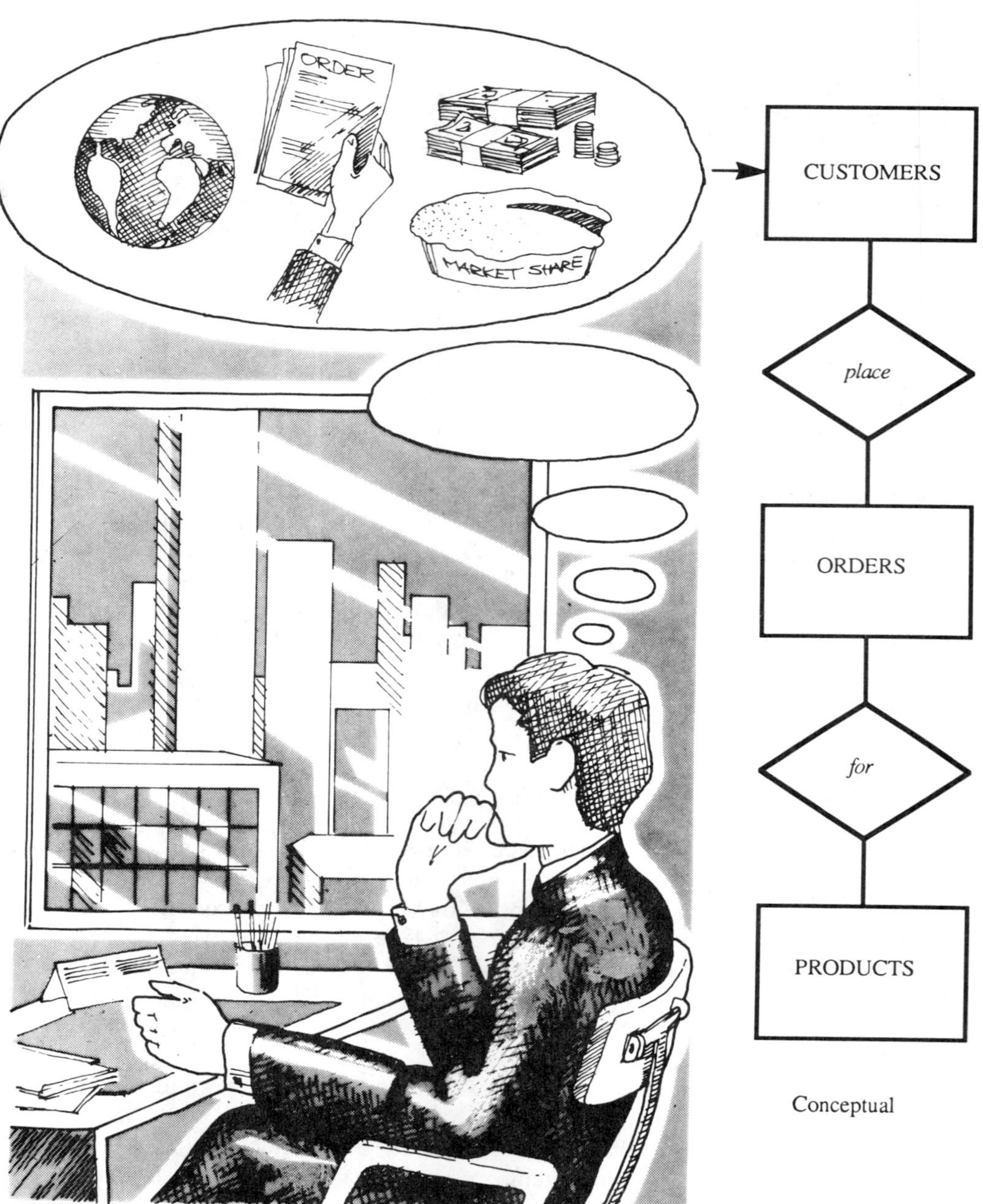

Figure 19: Three levels of data

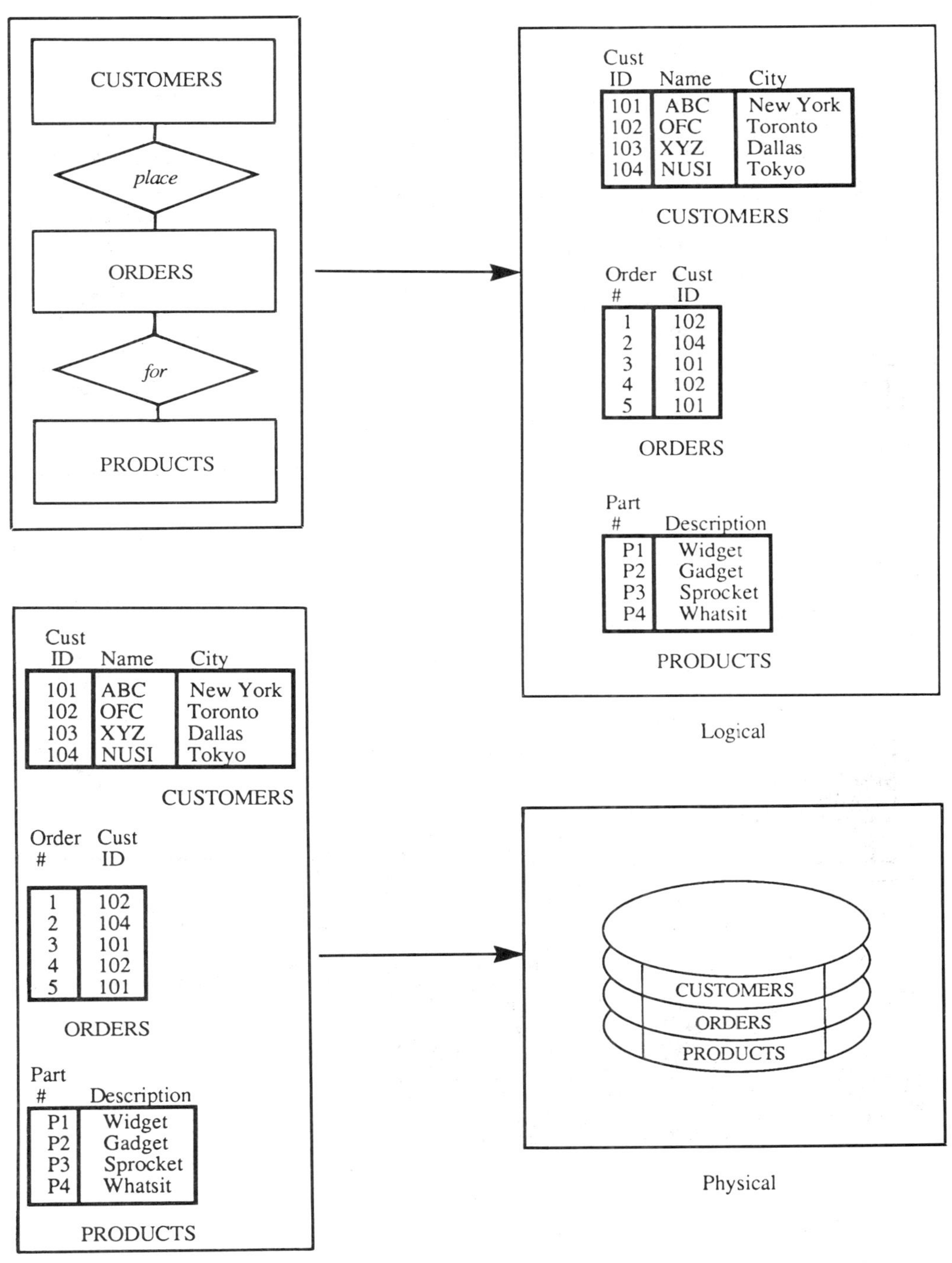
CUSTOMERS
place
ORDERS
for
PRODUCTS
Cust ID
Name
City
101 ABC New York
102 OFC Toronto
103 XYZ Dallas
104 NUSI Tokyo
CUSTOMERS
Order #
Cust ID
1 102
2 104
3 101
4 102
5 101
ORDERS
Part #
Description
P1 Widget
P2 Gadget
P3 Sprocket
P4 Whatsit
PRODUCTS
Logical
Cust ID
Name
City
101 ABC New York
102 OFC Toronto
103 XYZ Dallas
104 NUSI Tokyo
CUSTOMERS
Order #
Cust ID
1 102
2 104
3 101
4 102
5 101
ORDERS
Part #
Description
P1 Widget
P2 Gadget
P3 Sprocket
P4 Whatsit
PRODUCTS
CUSTOMERS
ORDERS
PRODUCTS
Physical

One also strives for *data independence.* This means that an application using a file or files should not depend on how the data are stored in the computer, and also that the addition or deletion of information in the files should not affect programs that do not use that information. Many applications today are *data dependent*: the knowledge of the data and its organization are built into the program logic. In applications that are data dependent, it is very difficult or impossible to change the storage organization (for example from index sequential to hash) without major modifications or complete rewrites of the software. Data independence is desirable because it is *always* easier and less costly to change the logic of a process than it is to change the data used by that process. But most importantly, the database or file system must serve the user and not the user serve the file system, which unfortunately is often the case.

For an analyst, database theory provides some fundamental knowledge regarding both database design and the functions that extract information from (or update) a database. One more comment before proceeding. Our goal for readers is not to make them database experts; this is its own area of specialization. The goal here is to cover basic database concepts and terminology, and to introduce some tools used in logical data modeling. These tools are useful for both traditional software development and for approaches using fourth-generation technologies.

4.3 Terminology and Concepts

In this section some commonly used database terminology and concepts, together with examples, are introduced. These will be used in the next section when, through a process called *normalization*, a procedure for deriving a well-behaved, logical file system is developed. (Throughout this chapter, the terms *logical model* and *relational model* are considered synonymous.)

One can think of a *logical database* or *file system* as being a level of abstraction lower than the conceptual model, and as a level of abstraction higher than the physical system (Figure 19). As with the conceptual model, one is not concerned with how the data are stored, but instead, with the overall information content. In the physical system, one is concerned with the storage of files on computer.

A logical database is made up of "logical" files, each composed of logically related records. Each record is composed of related items of information, called *attributes* or *fields*. Fields are to relations in the logical model what attributes are to entities in the conceptual model. A logical file gives information about some discernable aspect or entity of an enterprise. Entities and entity sets have been discussed in the previous chapter. Recall that entity sets for a company may include

- EMPLOYEES
- the PRODUCTS it sells
- the SUPPLIERS from whom products are purchased
- the CUSTOMERS who buy them
- the ORDERS that may result.

A logical record may or may not correspond to a physical record. For example, consider a computerized inventory file having a hash organization and keyed by a part number (Partnum). (A *key* or *primary key* is a single field or group of fields that uniquely identifies a record.) Assume that the following fields are contained in each record:

Partnum

Description

Supplier

Unit of Measure

Unit Cost

Quantity on Hand.

In addition, because of the hash organization, assume there are two pointer fields in each physical record: one that indicates the disk address of the previous record in key sequence and the other that indicates the disk address of the succeeding record in key sequence. These pointer fields are necessary because the physical sequence on disk does not correspond to the sequence of the key, and the facility for listing inventory items in key sequence is required. The physical record may appear as

Partnum	Description	Supplier	UoM	Cost	Qty	From	To

but the From and To fields containing the appropriate record addresses are not part of the logical record: that is, these fields do not contribute to the record's information content. Thus a logical record is not necessarily equal to a physical record.

A *relation* is a logical file in which all records have the same format and fixed length, and in which each record is unique (no duplicates). An alternative description is that a relation can be represented as a table or matrix having no duplicate rows. Because relations have a simple structure and are widely used in database design, our discussion is limited to them. In fact, almost all of the current research in databases concerns the relational data model. A relation, according to our definition, is also called a *first normal-form file.* A record or row is referred to as a *tuple*, and a column in the table is called an *attribute* or field.

The logical inventory file described above is a relation and is shown in table form in Figure 20.

Partnum	Description	Supplier	Unit of Measure	Unit Cost	Qty
127-74A	Note pads, 5" x 7"	ABC00	Dozen	6.25	23
A1432	Erasers, 1" x 3"	OFC00	100	9.85	420
BC801	Pencils, hb	OFC00	Gross	5.95	7
DAB-13A	Black cabinet	SMI10	Each	270.	2
FLEX785	Solvent, bulk	FLE50	Lb.	16.28	0
FEFC2A6	Black cabinet	EXE30	Each	658.	1
SX398	Paper clips	ABC50	Box	.38	62

Figure 20: INVENTORY

The file has the single-field or atomic key, Partnum. As with entity sets, some files may have more than one potential key. In the above example, if one assumes that a single supplier does not supply more than one type of the same item (i.e., Office Supply – OFC00 – offers only one type of HB pencil), then the inventory file in Figure 19 could also be keyed by the composite key Description-Supplier. If one did choose Description-Supplier as the key, then the fields Description and Supplier would be *subkeys*.

4.3.1 Relational Algebra

Relations are studied by database designers, and the manipulative operators they often use that act on relations, form what is called *relational algebra*. Relational algebra was originally defined by E.F. Codd at IBM, and discussion of relational algebra is derived from the text by C.J. Date [9].

Eight manipulative operations are defined, the first three of which readers may recall from mathematical set theory. These operations are as follows:

1. *union* forms a new relation from two given relations consisting of all records appearing in one or the other or both relations
2. *intersect* forms a new relation from two given relations consisting only of records appearing in both relations
3. *difference* forms a new relation from two given relations A and B, consisting of all records appearing in A and not in B
4. *select* creates a new relation by extracting specified records from a given relation
5. *project* creates a new relation by first extracting specified attributes from a given relation and then eliminating duplicates
6. *product* forms a relation from two given relations by concatenating all possible pairs of records, one from each of the given relations
7. *join* forms a new relation from two given relations by concatenating all possible pairs of records, one from each of the given relations, where the concatenation occurs based upon field values common to both relations
8. *divide* forms a relation from two given relations, one with two columns (the dividend), and the other with one column (the divisor), by extracting all values in one column of the dividend whose values in the other column match all values in the divisor.

Even though the eighth operator, *divide*, seems confusing, it is often useful to database designers. An overview of the first five operators appears in Figure 21, followed by examples of all eight operators.

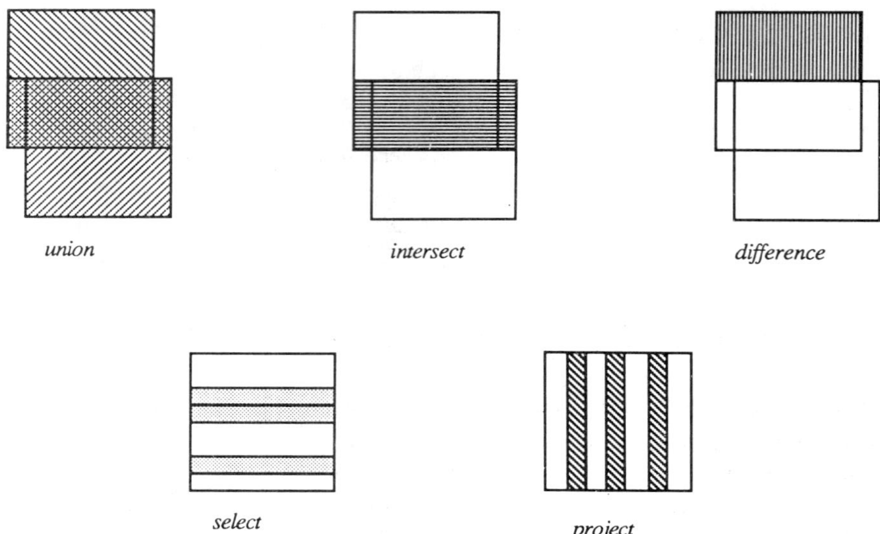

Figure 21: Overview of *union, intersect, difference, select, project*

Source: C.J. Date, AN INTRODUCTION TO DATABASE SYSTEMS, Volume 1, Fourth Edition, © 1986, Addison-Wesley Publishing Company, Inc., Reading, Massachusetts. pg. 259, Fig. 13.1. Reprinted with permission.

Notice that the result produced by each operation is a relation. This is the property of *closure*, and one can speak of relational algebra as being a closed system. Relational algebra defines the operations permitted within the relational data model.

4.3.1.1 The Operations union, intersect and difference

Consider a company with two branch locations, Oakville and North Burnside, and assume there are two supplier files, one for the Oakville branch and one for the North Burnside branch. Each file gives information concerning suppliers for the particular items at that location. These relations are shown in Figure 22, and the *union* of these two files is shown in Figure 23.

Supp-Id	Supplier Name
ABC00	ABCO, Inc.
ABC50	ABC Office Systems
OFC00	Office Supply

SUPPLIERS - OAKVILLE

Supp-Id	Supplier Name
ABC50	ABC Office Systems
FLE50	Flexible Equipment
OFC00	Office Supply

SUPPLIERS - NORTH BURNSIDE

Figure 22: Supplier files of Oakville and North Burnside

Supp-Id	Supplier Name
ABC00	ABCO, Inc.
ABC50	ABC Office Systems
OFC00	Office Supply

Figure 23: SUPPLIERS-OAKVILLE *union* SUPPLIERS-NORTH BURNSIDE

Notice that there are two records common to both relations, but they appear only once in the relation that results from their *union.* (Duplicates must be eliminated in order to comply with the definition of a relation.) In fact, these two common records also form the relation that is the intersection (Figure 24) of the two relations

SUPPLIERS - OAKVILLE
and
SUPPLIERS - NORTH BURNSIDE.

Supp-Id	Supplier Name
ABC50	ABC Office Systems
OFC00	Office Supply

Figure 24: SUPPLIERS-OAKVILLE *intersect* SUPPLIERS-NORTH BURNSIDE

There is only one record that appears in SUPPLIERS - OAKVILLE and not in SUPPLIERS - NORTH BURNSIDE:

ABC00 ABCO, Inc.

That record constitutes the relation SUPPLIERS - OAKVILLE *difference* SUPPLIERS - NORTH BURNSIDE. Of course the operators *union*, *intersect* and *difference* make sense only for two relations having the same format.

4.3.1.2 The Operations select, project, product, join and divide

Suppose a relation about employees includes the personnel number, title, name and salary of each individual.

Emp-Id	Title	Name	Salary
100	Trainee	JA Smith	12,000
266	Manager	JW Mosevich	55,000
293	Typist	RG Lawson	17,250
384	Programmer	JL Woeller	22,500
452	Manager	KA Zinszer	62,000
869	Trainee	LJ Denhart	16,000

EMPLOYEES

A new relation, MANAGERS, can be formed by using the *select* operator to extract those records where Title = Manager. The result is shown below.

Emp-Id	Title	Name	Salary
266	Manager	JW Mosevich	55,000
452	Manager	KA Zinszer	62,000

MANAGERS

Because this file consists entirely of managers, the field Title is unnecessary. This field can be eliminated by extracting the columns Emp-Id, Name and Salary, using the *project* operator.

Emp-Id	Name	Salary
266	JW Mosevich	55,000
452	KA Zinszer	62,000

PROJECT MANAGERS(Emp-Id, Name, Salary)

The *product* operator is similar to the notion of the Cartesian product from mathematics. Suppose the company under consideration sells binders in various colors and sizes for note paper. Information about the colors and sizes is given in the two relations BINDER-COLOR and BINDER-SIZE, respectively.

Color
Black
Blue
Green
Red
White

BINDER-COLOR

Size
9 x 12
5 x 7
4 x 6

BINDER-SIZE

Then the *product* of these two relations, BINDER-COLOR TIMES BINDER-SIZE, is the set of all (binder-color, binder-size) pairs.

Color	Size
Black	9 x 12
Black	5 x 7
Black	4 x 6
Blue	9 x 12
Blue	5 x 7
Blue	4 x 6
Green	9 x 12
Green	5 x 7
Green	4 x 6
Red	9 x 12
Red	5 x 7
Red	4 x 6
White	9 x 12
White	5 x 7
White	4 x 6

BINDER-COLOR TIMES BINDER-SIZE

To illustrate *join* recall the INVENTORY file shown in Figure 20 which shows each Part Number, Description, Supplier, Unit of Measure, Unit Cost and Quantity of inventory. Using the *project* operator on the attributes Partnum, Description, Supplier in the INVENTORY relation, a new relation, called PRODUCT, can be formed (Figure 25).

Partnum	Description	Supplier
127-74A	Note pads, 5" x 7"	ABC00
A1432	Erasers, 1" x 3"	OFC00
BC801	Pencils, HB	OFC00
DAB-13A	Black cabinet	SMI10
FLEX785	Solvent, bulk	FLE50
SEFC2A6	Black cabinet	EXE30
SX398	Paper clips	ABC50

Figure 25: PRODUCT

Suppose a relation called SUPPLIER contains information about suppliers of products (Figure 26).

Supp-Id	Supplier Name	Supplier Street
ABC00	ABCO, Inc.	573 Barton
ABC50	ABC Office Systems	95-A Union
EXE30	Executives, Inc.	900 North
FLE50	Flexible Equipment	5 Park
OFC00	Office Supply	1000 East Third
SMI10	Smith's Service	194 Indian Park
ZEN00	Zenith Unlimited	934 Queen

Figure 26: SUPPLIER

Notice there is a field, the supplier's identifier (called Supplier in the relation PRODUCT and Supp-Id in the relation SUPPLIER) common to both files. They *join* using that common field to form the relation PRODUCT-SUPPLIER (Figure 27). (Notice that the result has considerable redundancy, and that the order of the attributes and tuples is immaterial.)

Partnum	Description	Supp-Id	Supplier Name	Supplier Street
127-74A	Note pads, 5" x 7"	ABC00	ABCO, Inc.	573 Barton
A1432	Erasers, 1" x 3"	OFC00	Office Supply	1000 East Third
BC801	Pencils, HB	OFC00	Office Supply	1000 East Third
DAB-13A	Black cabinet	SMI10	Smith's Service	194 Indian Park
FLEX785	Solvent, bulk	FLEX785	Flexible Equipment	5 Park
SEFC2A6	Black cabinet	EXE30	Executives, Inc.	900 North
SX398	Paper clips	ABC50	ABC Office Systems	95-A Union

Figure 27: PRODUCT-SUPPLIER

One supplier, Zenith Unlimited, has no corresponding record in PART, and does not participate in the *join* operation.

To conclude the discussion of relational algebra, we consider the *divide* operator and return to the company with two branch locations, one in Oakville and the other in North Burnside. A relation about suppliers and locations can be composed by combining the relations in Figure 22 to form a new relation, SUPPLIER-LOCATION.

Supp-Id	Location
ABC00	Oakville
ABC50	Oakville
FLE50	North Burnside
OFC00	North Burnside
OFC00	Oakville

SUPPLIER-LOCATION

This relation describes which suppliers supply to what locations, and forms the *dividend* of the operator. Suppose the relation LOCATION contains the list of locations.

Location
North Burnside
Oakville

LOCATION

We *divide* SUPPLIER-LOCATION by LOCATION, the result is a list of those suppliers who supply to both locations, as shown below.

Supp-Id	Location		Location		Supp-Id
ABC00	Oakville		North Burnside		ABC50
ABC50	Oakville	÷	Oakville	=	OFC00
ABC50	North Burnside				
FLE50	North Burnside				
OFC00	Oakville				
OFC00	North Burnside				
SUPPLIER-LOCATION (dividend)		÷	LOCATION (divisor)	=	SUPPLIERS OF BOTH LOCATIONS (quotient)

Relational algebra defines the operations permitted on a set of relational files. These operations permit the building of a dynamic database because of the property of closure. The operations may be used for both traditional software development, and for development using fourth-generation technologies. In fact, in Chapter 11, the NOMAD language equivalents of relational algebra are explored.

4.3.2 Relationships in the Logical Model

Recall from the discussion of the conceptual model that a relationship was defined to be an association among entities. Relationships are defined in the relational model as well.

Consider again the two relations SUPPLIER and INVENTORY (Figures 26 and 20, respectively). For each value of the attribute Supp-Id in SUPPLIER, there may be several records in INVENTORY with that same value for the attribute Supplier. For example, OFC00, Office Supply, supplies both 1″ x 3″ erasers and HB pencils. However, for each value of Supplier in INVENTORY, there is only one corresponding record in SUPPLIER, namely, the supplier of that particular part. FLE50 supplies bulk solvent and corresponds to the record in SUPPLIER whose key has the value FLE50.

There is a *one-to-many relationship* (1:N) between the records in the relation SUPPLIER and the records in INVENTORY. One can envision a heirarchy and refer to the records in SUPPLIER as the *parent* records, and to the records in INVENTORY as the *child* records. This relationship is illustrated by the following *data structure diagram.*

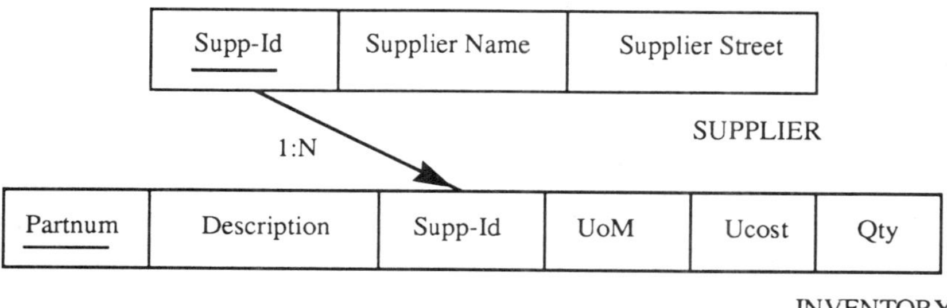

The direction of the arrow indicates the direction of the relationship. The relationship between these relations is permitted by the common field, Supp-Id, called the *connection field.*

Recall an example of earlier sections, the company with two branch locations, North Burnside and Oakville. Suppose the relation called PRODUCT shown in Figure 25 provides information concerning the items offered for sale, and the relation BRANCH shown below contains information about the two branches.

Location	Employees
North Burnside	57
Oakville	89

BRANCH

An additional relation providing information about quantities of stock at the two locations might look something like this:

Partnum	Location	Qty
127-74A	North Burnside	20
127-74A	Oakville	3

A1432	North Burnside	0
A1432	Oakville	14
BC801	North Burnside	234
BC801	Oakville	12
.	.	.
.	.	.
.	.	.

STOCK.

For each record in PRODUCT, there may be several records in STOCK; also, for each record in BRANCH, there may be several records in STOCK. But for a given STOCK record, there is only one corresponding record in PRODUCT, and likewise for BRANCH. So both PRODUCT and BRANCH participate in one-to-many relationships with STOCK. These relationships can be shown in a data structure diagram.

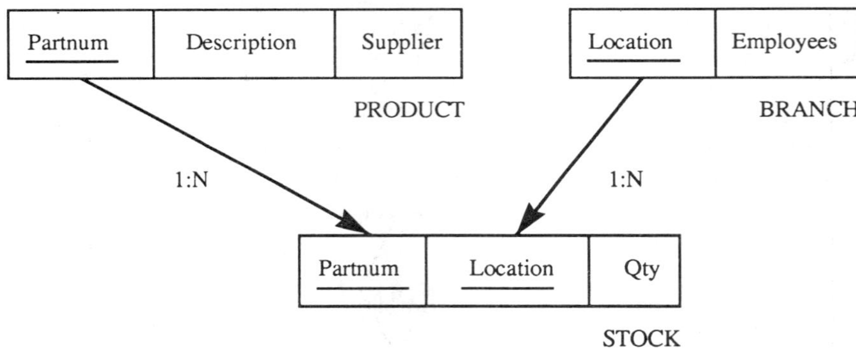

In the database shown above, there are two 1:N relationships: between PRODUCT and STOCK, and between BRANCH and STOCK. There is also a relationship between PRODUCT and BRANCH, but it is of a different type. At each location, there are many different products, characterized by all the different part number codes (Partnum values) at that location. Similarly, one product may occur in all (both) locations. There is a *many-to-many (N:M) relationship* between PRODUCT and BRANCH.

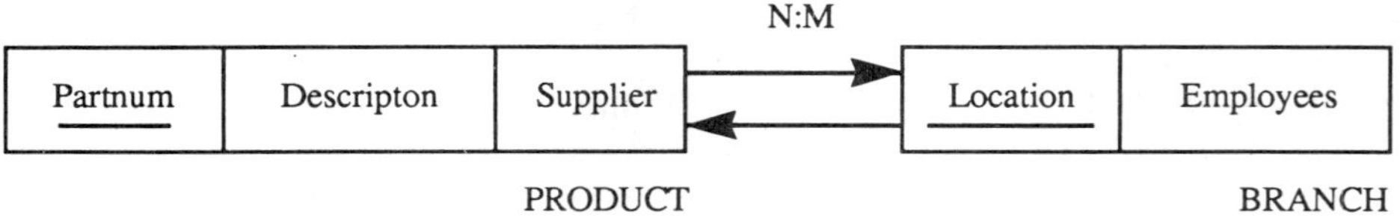

In general, for two relations A and B between which there is a many-to-many relationship, this relationship can always be decomposed into two 1:N relationships by building a *bridge* between them. The bridge in the above example is STOCK. The more general case is shown below.

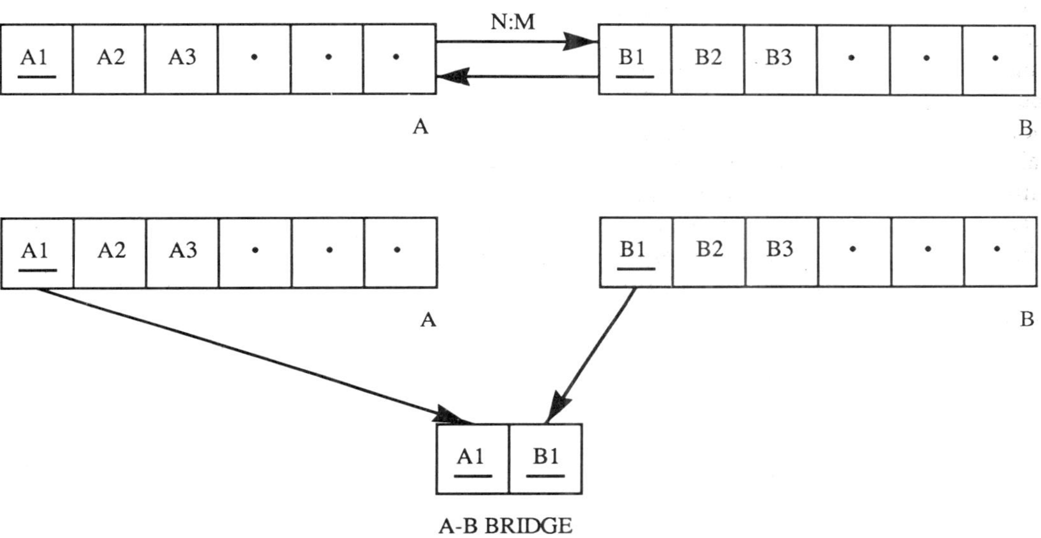

The key for the bridge is the concatenated keys of A and B. Because any two files participating in a many-to-many relationship can be decomposed via a bridge into two one-to-many relationships, discussion is limited to one-to-many relationships. (Notice how similar this is to the relationships covered in discussions of the conceptual model.)

4.3.3 Functional Dependencies in Logical Files

We now reintroduce another concept necessary for the derivation of the logical data base: the notion of *functional dependency* as it applies to attributes within a relation.

The notions of functions and dependencies were introduced within the context of the E-R model. These concepts are appropriate for the relational model as well. For example, consider Burnside University, and suppose there are two faculties, Arts and Science. The English, Drama, French, and Fine Arts departments are in the Arts faculty. Mathematics, Geology, Physics and Biology comprise the Science faculty. Then the set of DEPARTMENT elements is English, Drama, French, Fine Arts, Mathematics, Geology, Physics and Biology; the set of FACULTY elements is Arts and Science. For each element in DEPARTMENT, its faculty is uniquely determined. An alternate way of saying this is that FACULTY is *functionally dependent* on DEPARTMENT. This is illustrated in Figure 28.

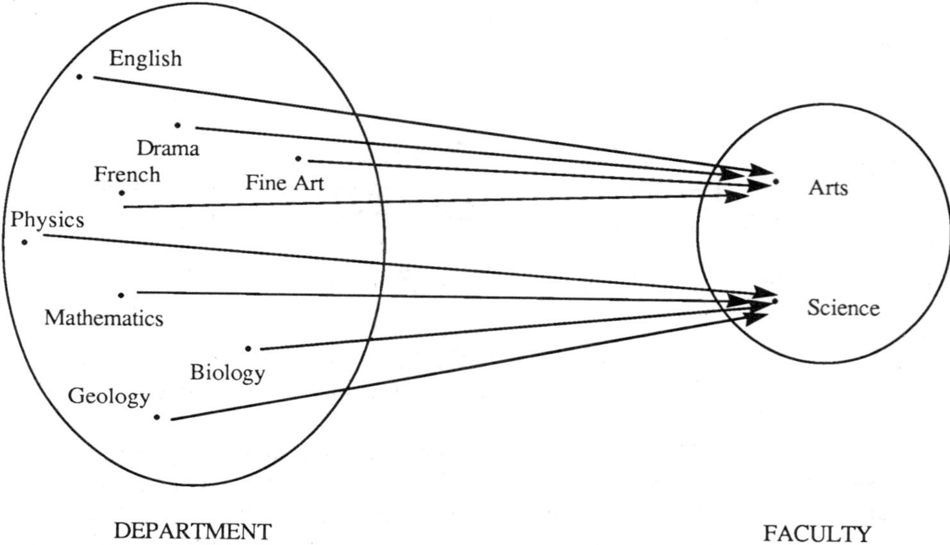

Figure 28: Functional dependency

Functional notation from mathematics is again used and one can write

f:DEPARTMENT ----> FACULTY.

Notice that there is a one-to-many correspondence between the elements of FACULTY and DEPARTMENT, because for any member of DEPARTMENT, English, for example, there is exactly one element in FACULTY that corresponds – the faculty to which that department belongs (Arts in this case). However, for any FACULTY member, for example Science, there are several DEPARTMENT members that correspond – all the departments in that Faculty. One could even go so far as to graph this function (Figure 29).

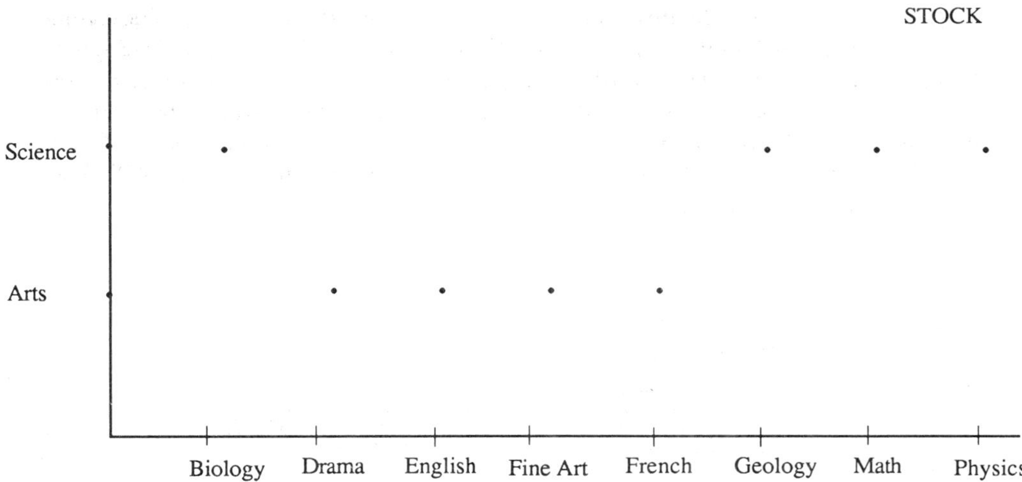

Figure 29: Graph of the function f: DEPARTMENT --> FACULTY

(It will be discovered in the next section that some functional dependencies are desirable and that others are undesirable and should be eliminated.)

Recall the file STOCK from the previous section, shown again in Figure 30. Remember that there are two branch locations, North Burnside and Oakville, and that STOCK provides information about the inventory of products at each location. It should be clear that the field Qty is functionally dependent on the entire key Partnum-Location, since for a given (part number, location) pair, the quantity is uniquely determined. Furthermore, the quantity cannot be determined when given only a value for Partnum, say BC801, because that (and any) product may occur at more than one location. Likewise, given a value for Location, the quantity cannot be determined. One can write

f:(Partnum, Location) ----> Qty.

Partnum	Location	Qty
127-74A	North Burnside	20
127-74a	Oakville	3
A1432	North Burnside	0
A1432	Oakville	14
BC801	North Burnside	234
BC801	Oakville	12
⋮	⋮	⋮

Figure 30: STOCK

This concept is expressible in terms of the notation and discussion developed thus far in the chapter, for if a functional dependency occurs among attributes in a file, then for a given value of the independent attribute (or attributes), the value of the dependent attribute is completely determined. This has already been illustrated with the example of the departments and faculties of Burnside University.

Department	Faculty
Biology	Science
English	Arts
Drama	Arts
French	Arts
Fine Arts	Arts
Geology	Science
Mathematics	Science
Physics	Science

DEPARTMENTS

Recall from set theory and from previous discussions regarding functions that we are concerned with unique values only, and duplicates are disregarded. Within the context of relations and relational algebra, the *project* operator can be used to extract the unique values of Faculty. The following data structure diagram shows the association of those unique values to the corresponding values in DEPARTMENT and illustrates the dependency of Faculty on Department.

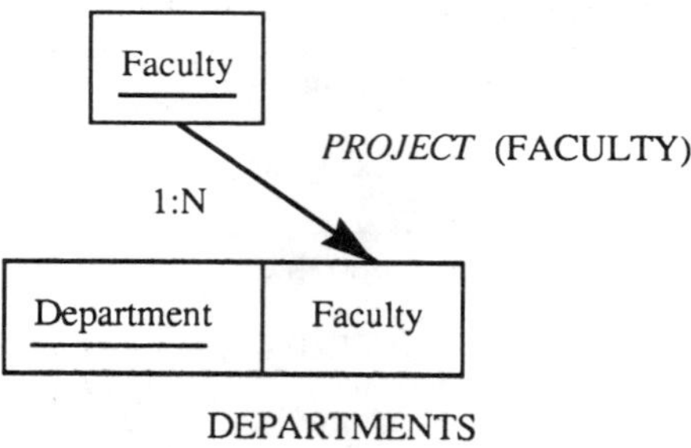

DEPARTMENTS

This fairly intuitive presentation of functional dependencies concludes with three remarks. First, this section has reintroduced the concept of functional dependency with respect to non-mathematical values, and the notion of a one-to-many correspondence between attributes *within the same file*, whereas the previous chapter was concerned with one-to-many relationships *between different entities*. Second, if a one-to-many relationship occurs between corresponding (unique) attribute values within a file, one suspects that a functional dependency may occur, that is, the one-to-many relationship is a consequence of the dependency. The reverse, however, may not be true: there may be one-to-many relationships that are merely coincidences, and not functional dependencies. There are commercially available software products that can detect one-to-many relationships (the symptoms) between attributes; however, only a human being can provide the diagnosis.

Finally, a complete treatment of the topic of functional dependency is beyond the scope of this text, and those readers who are interested in more material may refer to James Bradley's text, *Introduction to Data Base Management in Business* [4].

4.4 Procedures for Developing a Relational Database

So far in this chapter the concepts of relations, relational algebra, relationships between relations and the notion of functional dependencies within a file have been introduced. In this section these ideas are combined to give a procedure for deriving a logical database. To reiterate, the objective here is to produce a file system that will serve the user community, be as simple as possible, and be free from anomalies such as the possibility of inconsistent updating. To arrive at such a desirable result, our starting point is the corporate data model. The entities are first of all translated into relations. Then, by placing increasing restrictions on the permissible dependencies, one arrives at a desirable end product. The resulting relations are called *higher normal-form files*.

4.4.1 Transforming Entities into Relations

The place to begin in the process of deriving a logical data model or database is the previously derived conceptual data model. The entity sets of the E-R model form the basis for the logical database. Recall that a relation can be represented as a table having no duplicate rows. The translation or decomposition process then, from the conceptual to the logical level, requires that the entities resemble tables, as in Figure 31.

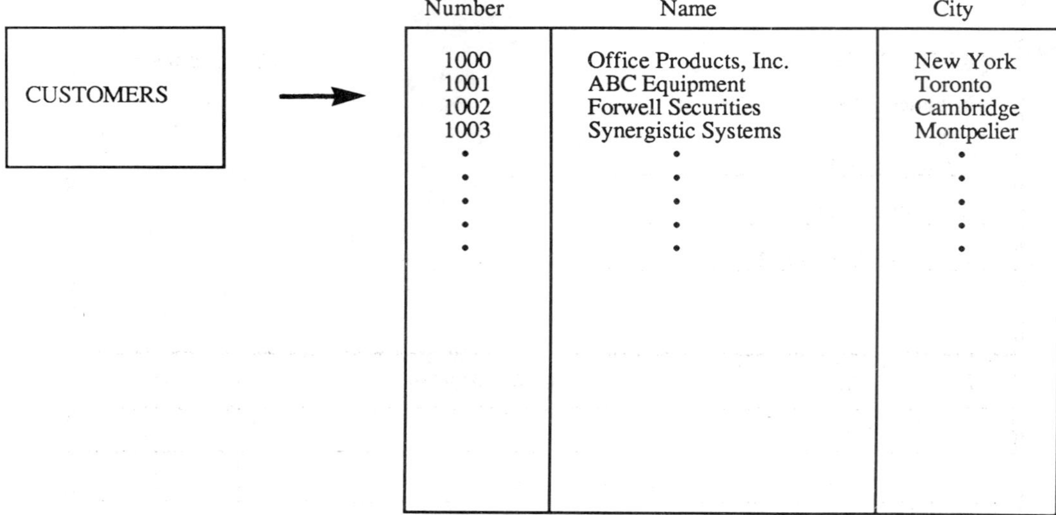

Figure 31: The entity CUSTOMERS and the table CUSTOMERS

However, not all entities may have this rather restricted form, for consider the entity ORDERS in the E-R diagram shown in Figure 32.

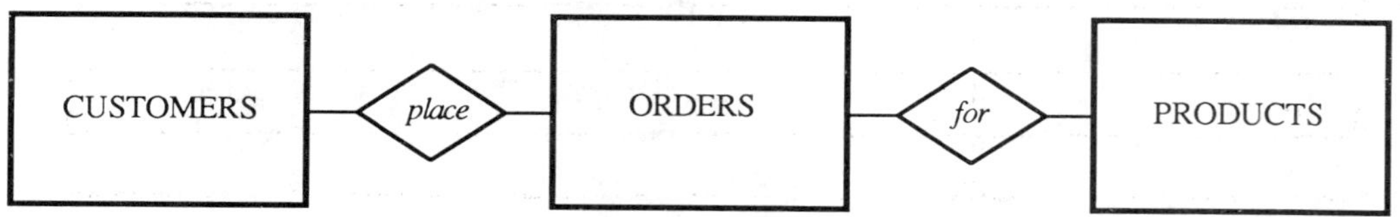

Figure 32: E-R diagram of CUSTOMERS *place* ORDERS *for* PRODUCTS

Individual orders from customers are members of the entity ORDERS. An order from one customer, order number 12573, appears in Figure 33. An order from another customer, order number 12574, appears in Figure 34.

SALES ORDER

ORDER: 12573 DATE: June 30

Customer: 1000

Office Products

123 Fifth Avenue

New York, NY 10012

LINE	ITEM	DESCRIPTION	QUANTITY
1	127-74A	NOTE PAD, 5" x 7"	10
2	A1432	ERASERS, 1" x 3"	5
3	BC801	PENCILS, HB (BOX)	3
4	DAB-13A	BLACK CABINET, 15" x 24" x 36"	1
5	FLEX785	SOLVENT, BULK (POUND)	3
6			
7			
8			
9			
10			
11			

Figure 33: Sample order from customer 1000

Notice the differences in the two orders. Customer 1000 has ordered five items and customer 1001 has ordered only one item. One may conclude generally that ORDERS is not a relation because individual orders may vary as to length and format.

SALES ORDER

ORDER: 12574 DATE: June 30

Customer: 1001

ABC Equipment

5100 Yonge Street

Toronto, ON M5W 2F8

LINE	ITEM	DESCRIPTION	QUANTITY
1	5X398	PAPERCLIPS (BOX)	4
2			
3			
4			
5			
6			
7			
8			
9			
10			
11			

Figure 34: Sample order from customer 1001

This problem may be overcome, however, by decomposing (or exploding) the entity ORDER as shown in Figure 35. The entity ORDER has been separated or decomposed into two entities:

- ORDER corresponding to the overall information on the order
- ORDER LINES corresponding to individual lines on the order.

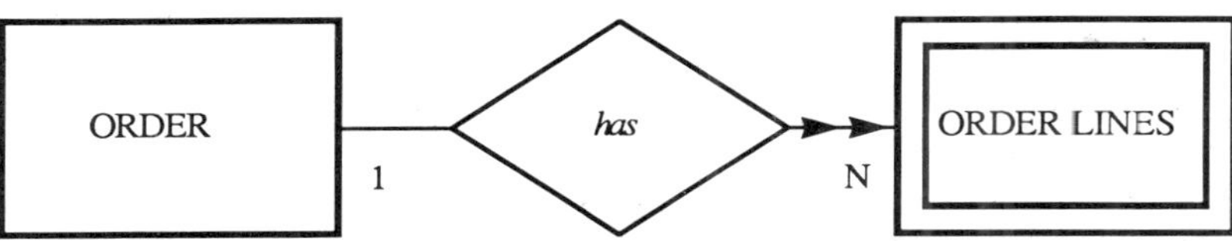

Figure 35: Explosion of ORDER

Notice the double arrow between ORDER and ORDER LINE pointing toward ORDER LINE. This is meant to imply that the entity set ORDER LINE is not only existence dependent on the entity set ORDER, but also that it must be associated with a particular order. Within this context, the attributes pertaining to each entity can be assigned to either ORDER or ORDER LINE, as appropriate (Figure 36).

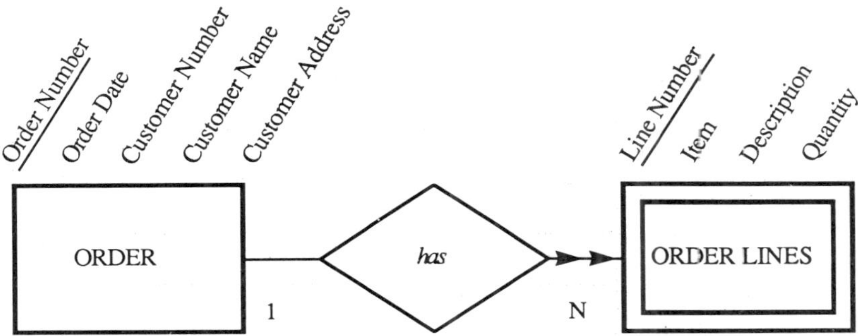

Figure 36: E-R diagram of ORDER, ORDER LINES

It is now possible to represent both entities as relations (Figure 37). The constraint of the double arrow in the E-R diagram implies that the key of ORDER LINES is obtained by concatenating the key of the pertinent order and the line number of that order. Alternatively, this can be shown in data structure diagram form.

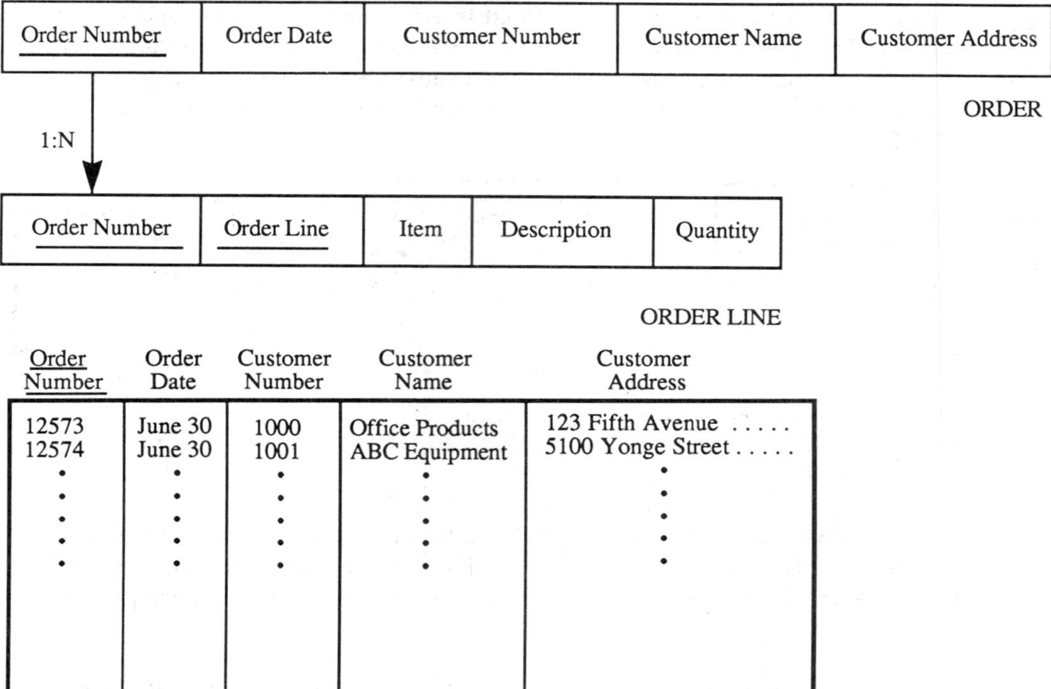

Order Number	Order Date	Customer Number	Customer Name	Customer Address
12573	June 30	1000	Office Products	123 Fifth Avenue .....
12574	June 30	1001	ABC Equipment	5100 Yonge Street .....
•	•	•	•	•
•	•	•	•	•
•	•	•	•	•
•	•	•	•	•
•	•	•	•	•

ORDER

Order Number	Line Number	Item	Description	Quantity
12573	1	127-74A	NOTE PAD, 5" x 7 "	10
12573	2	A1432	ERASERS, 1 " x 3'	5
12573	3	BC801	PENCILS, HB(BOX)	3
12573	4	DAB-13A	BLACK CABINET 15" x 24" x 36"	1
12573	5	FLEX785	SOLVENT, BULK (POUND)	3
12574	1	5X398	PAPERCLIPS (BOX)	4
•	•	•	•	•
•	•	•	•	•
•	•	•	•	•
•	•	•	•	•
•	•	•	•	•
•	•	•	•	•

ORDER LINE

Figure 37: ORDER, ORDER LINES as relations

To users, such a decomposition (of ORDER into the two entity sets ORDER and ORDER LINE) probably represents a technical detail. It is only for the purpose of translating to the relational model that this is done. Entities requiring such explosion (to facilitate translation into relations) may be shown as in Figure 38.

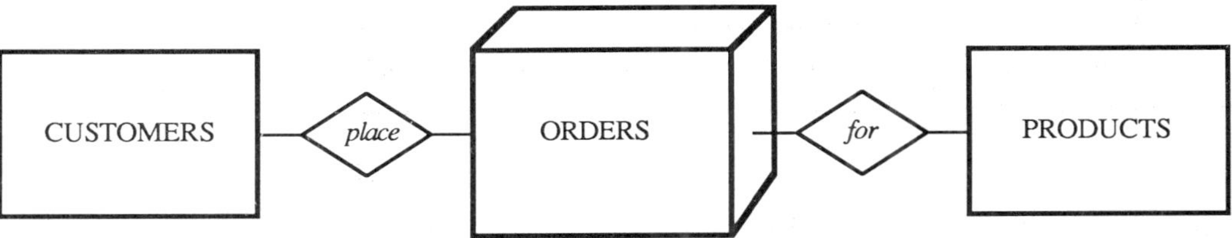

Figure 38: Entities requiring decomposition

The attempt to represent ORDERS three dimensionally is meant to illustrate the fact that ORDERS has some things "inside" that require technical analysis. A separate drawing is sometimes used for showing the decomposition or "interior" of an entity such as ORDERS.

Consider another example: Figure 14 depicts a simplified university model of STUDENTS, SEMESTERS, COURSES, DEPARTMENTS and so on. Students take various courses in different semesters. Suppose that English 102 is offered in the fall and winter semesters. During these two semesters, the class is given at two different times: section 1 meets at 10:00, and section 2 meets at 2:00. This information cannot be translated directly to the relational model, because the class is given twice. However, through the process of explosion, the entities can be transformed into relations (Figure 39). A relation is also called a first normal-form (1NF) file.

The rest of this section is devoted to developing the procedure for arriving at a desirable logical structure. All the dependencies between fields or attributes in a file will be determined, and those that are undesirable will be eliminated. The primary objective is to derive a set of relations in which all fields are functionally dependent on exactly the *primary key* (the field or fields that uniquely identify a record).

4.4.2 Second Normal-Form (2NF) Files

The first restriction placed on a set of relations is that they satisfy the following definition.

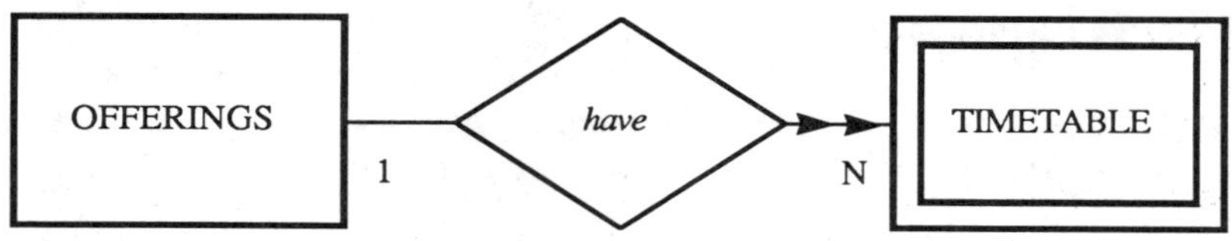

Course	Semester
English 102	Fall
English 102	Winter
French 100	Fall
French 101	Winter
French 102	Summer
⋮	⋮

OFFERINGS

Course	Semester	Section	Time
English 102	Fall	01	10:00
English 102	Fall	02	2:00
English 102	Winter	01	10:00
English 102	Winter	02	10:00
French 100	Fall	01	8:00
French 101	Winter	01	9:00
French 102	Summer	01	10:00
⋮	⋮	⋮	⋮

TIMETABLE

Figure 39: OFFERINGS *have* TIMETABLE

A *second normal-form (2NF) file* is

1. a 1NF file
2. such that each nonkey field is functionally dependent on the whole key, but not on any subkey fields if the key is composite (composed of more than one field).

Note that all 1NF files with atomic or single-field keys are automatically 2NF files. Two examples dealing with composite keys follow.

Consider from previous discussions the 1NF file called STOCK shown in Figure 30. It has the composite key Partnum-Location. The nonkey field Qty clearly satisfies the definition of 2NF, so the file is in second normal form. Suppose the attribute Description is added to the file and the result is called STOCKED PRODUCTS. It is shown below in both data structure diagram and table form.

Partnum	Location	Description	Qty

STOCKED PARTS

Partnum	Location	Description	Qty
127-74A	North Burnside	Note Pads, 5″x 7″	20
127-74A	Oakville	Note Pads, 5″x 7″	3
A1432	North Burnside	Erasers, 1″x 3″	0
A1432	Oakville	Erasers, 1″x 3″	14
BC801	North Burnside	Pencils, HB	234
BC801	Oakville	Pencils, HB	12
.	.	.	.
.	.	.	.
.	.	.	.

STOCKED PRODUCTS

What effect does the addition of the attribute Description have on this file? It is still in 1NF, and it is obvious that Description is functionally dependent on the subkey Partnum and *not* dependent on Location. This makes the file in

1NF but not in 2NF. Why is this an undesirable condition? First, on an intuitive level, the attribute Description does not relate directly to the entity regarding quantities of stock; it relates more to the products offered for sale. Secondly, redundancy has been introduced into the file, as well as the possibility of inconsistent updating. Suppose that Office Supply, the suppliers of part BC801, HB pencils, decide to change this product. The part number remains the same, but now the description reads Mechanical pencils. This description must be changed wherever it occurs in the file, and it is possible that because of a programming error one could in fact change the first occurrence but not the second. The result is inconsistent information contained in the two records

BC801	North Burnside	Mechanical Pencils	234
BC801	Oakville	Pencils, HB	12

One description is incorrect, and unless one has incorporated the checking of Description against Partnum in some programming logic, this sort of error may occur. The data have lost integrity – the file contains inconsistent/ambiguous data, and can convey no information for part number BC801.

This (potentially) undesirable condition is easily remedied by decomposing STOCKED PRODUCTS into two 2NF relations, one about the entity PRODUCTS and the other about STOCK.

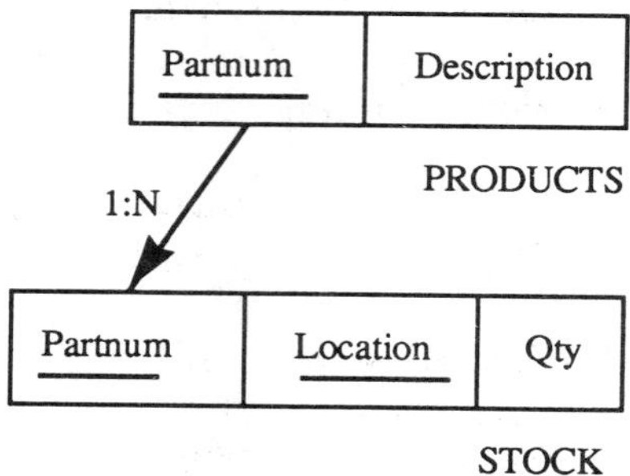

The process of deriving a "good" relational structure generally involves the elimination of undesirable dependencies by decomposing a given relation into two or more relations having only permitted dependencies.

A serious problem with 2NF files is that dependencies between nonkey fields are permitted; the definition of third normal form disallows this condition.

4.4.3 Third Normal-Form (3NF) Files

A *third normal-form (3NF) file* is one that is

1. a 2NF file

2. such that no nonkey fields are functionally dependent on any other nonkey fields.

Several examples follow. Assume a relation called CUSTOMERS contains information about customers to whom products are sold. Salespeople are assigned to customers according to this policy:

customer names beginning with

A through F have salesperson J.J. Smyth

G through M have salesperson B.J. Campbell

N through S have salesperson T.L. Nichols

T through Z have salesperson R.B. Simpson.

An excerpt from such a relation might appear as

Cust-Name	City	Sales-person	Purchases Last Year
ABCD Company	Cambridge	JJS	22,505
DEVCO, Inc.	Centerville	JJS	5,957
Markham Enterprises	Markham	BJC	197,900
New Deal Corp.	Plattsville	TLN	588
Uptown Downtown	Waterloo	RBS	58,939
.	.	.	.
.	.	.	.
.	.	.	.

CUSTOMERS.

Each nonkey field is functionally dependent on the key Cust-Name and no nonkey field depends on any other nonkey field. This file, then, is in third normal form.

For the next example, return to Burnside University. Suppose a relation INSTRUCTOR contains information about faculty members, part of which appears below.

Name	Department	Faculty
JC Berry	Fine Arts	Arts
LG Bourdon	Mathematics	Science
ML Crosby	Drama	Arts
WC Fuller	Physics	Science
FM Moser	Physics	Science
JE Prime	Biology	Science
VJ Walker	Drama	Arts
. .	.	.
. .	.	.
. .	.	.

INSTRUCTORS

The functional dependency of Faculty on Department has previously been considered, so this file is in 2NF but not in 3NF. Again, this is an undesirable condition because of the reasons discussed previously. On an intuitive level, the attribute Faculty does not relate directly to the entity INSTRUCTORS. As well, redundancy has been introduced into the file by the addition of the attribute Faculty. Finally, the possibility of inconsistent information appearing in the file has been introduced. Suppose that the Mathematics department will be changed from Science to the Arts faculty. Each instance of Mathematics must have its Faculty value changed to Arts instead of Science, and as previously stated, programming errors might result in some but not all occurrences being changed. The result would be that some Mathematics instructors may be in the Arts faculty and some may be in Science. In addition, another problem could also arise: if a new department were created, Information Science, but no instructors were as yet hired, there would be no way of storing such information.

A solution to these shortcomings is to decompose the relation INSTRUCTORS into two 3NF relations, one about departments and their corresponding faculties, and the other about instructors and their departments.

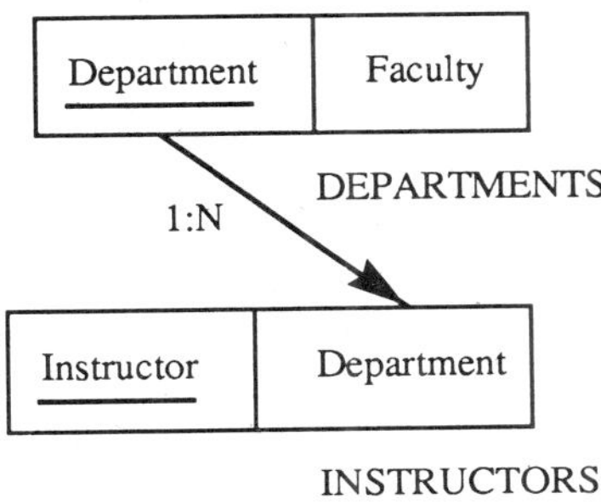

Another aspect of converting relations to 3NF is that all fields that can be calculated from other fields are eliminated. For example, a relation INVENTORY might contain information about the value of on-hand stock:

Item Number	Description	Qty	Unit Cost	Value
P1	Widget	10	2.50	25.00
P2	Sprocket	6	10.80	64.80
P3	Gasket	0	90.30	0

INVENTORY

Value is a function of Qty and Unit Cost, because for each Item Number in the relation,

Qty x Unit Cost = Value.

Then Value need not appear as data at all, but rather as part of a process. This restriction applies here to attributes within the same file. However, this process can be extended to consider attributes from among *different files*; attributes derivable from processes may be eliminated.

Item Number	Description	Qty	Unit Cost
P1	Widget	10	2.50
P2	Sprocket	6	10.80
P3	Gasket	0	90.30

3NF INVENTORY

It was believed, for a time, that 3NF files were free of undesirable complications. However, examples were uncovered in which subkey fields were dependent on nonkey fields, illustrated by the following example adapted from the text by James Bradley [4].

Recall once again the relation STOCK shown in Figure 30. Stock provides information about the products at each of two branch locations, North Burnside and Oakville. Suppose that at each location employees are responsible for certain parts. They reorder them when the on-hand quantities become low, check incoming shipments and so on. An employee may be responsible for several parts.

Partnum	Location	Employee
127-74A	North Burnside	JA Jackson
127-74A	Oakville	RE Taylor
A1432	North Burnside	KA Goetz
A1432	Oakville	AM Harris
BC801	North Burnside	KA Goetz
BC801	Oakville	RE Taylor
.	.	.
.	.	.
.	.	.

RESPONSIBILITY

Clearly the relation RESPONSIBILITY is in 3NF: given a part number-location pair, the employee is uniquely determined (the one who looks after that part at that location). Also, given either a specific part number or specific location, one cannot determine the associated employee. However, if one assumes that an employee works at one location only (which seems reasonable), then given an employee, the employee's location is determined, so Location is functionally dependent on Employee. This type of undesirable possibility was discovered by Boyce and Codd [7], and led to the definition that follows.

4.4.4 Boyce-Codd Normal-Form (BCNF) Files

A *Boyce-Codd Normal-Form (BCNF) file* is one that is

1. a 3NF file and

2. such that all subkey fields are functionally dependent on only the entire key.

Note that 3NF atomic-key files are automatically BCNF files.

The example file above, RESPONSIBILITY, was shown to be in 3NF but not in BCNF because the subkey Location is functionally dependent on the nonkey Employee. This is undesirable, for all of the reasons cited before: it goes against one's intuition regarding attributes and entities, provides for the possibility of inconsistent information being introduced into the file, and creates an inability to store certain information. RESPONSIBILITY can be decomposed into two BCNF relations, one containing information about employees and their locations, and the other about employees and the parts for which they are responsible.

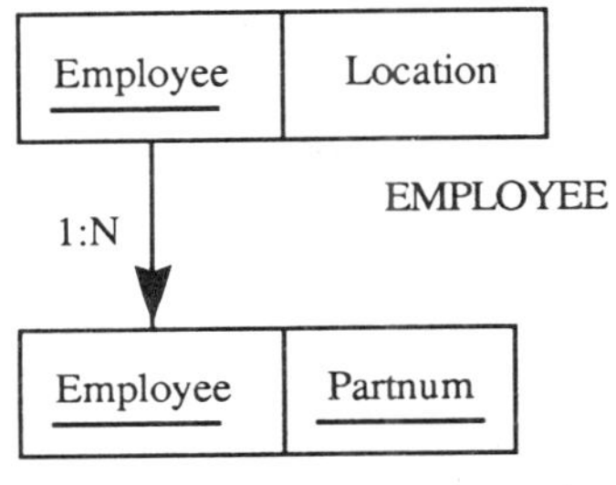

Employee	Location
KA Goetz	North Burnside
AM Harris	Oakville
JA Jackson	North Burnside
RE Taylor	North Burnside
.	.
.	.
.	.

EMPLOYEE

Employee	Partnum
KA Goetz	A1432
KA Goetz	BC801
AM Harris	A1432
JA Jackson	127-74A
RE Taylor	127-74A
RE Taylor	BC801
.	.

EMPLOYEE-PRODUCT

4.4.5 Fourth and Fifth Normal-Form Files

Surely Boyce-Codd Normal Form is the simplest possible. Unfortunately this is not the case, and in 1977 R. Fagin [11] discovered a more elaborate dependency called a *multivalued dependency*. These dependencies involve many-to-many (N:M) relationships among attributes of a file. They are, in fact, a more general case of functional dependencies (1:N relationships), and permit certain insertion and deletion anomalies in databases. Insertion of one record usually requires insertion of additional records, and deletion of a record usually requires other records to be deleted. BCNF files permit such dependencies, and so the definition of fourth normal-form files follows.

A *fourth normal-form (4NF) file* is one that is

1. a BCNF file

2. such that each multivalued dependency is also a functional dependency.

Fortunately such cases are relatively rare, yet they do exist. An example of a relation that contains a multivalued dependency follows.

Consider an automobile manufacturer who makes two models, the Ordinaire and the Connoisseur. For each model there are several sets of options to choose from, and the purchaser is offered luxury "extras." Suppose the options for the Ordinaire are called Dull and Boring, and the options for the Connoisseur are called Fancy, Deluxe and Super. There are dealers throughout the country, and each dealer sells only one model; however, each dealer offers all options of the model.

Model	Option	Dealer
Ordinaire	Dull	Good Cars Inc.
	Boring	Ordinaire Dealer
		Orville Motors
Connoisseur	Fancy	Century Auto
	Deluxe	Connoisseur Cars
	Super	

The models are shown in the left-hand column, the options for each model are shown in the middle and the dealers are shown on the right. Notice that this file is not a relation. Such files are sometimes called *unnormalized*. A normalized file called AUTOS is constructed to hold this information (Figure 40).

Model	Option	Dealer
Connossieur	Deluxe	Century Auto
Connossieur	Deluxe	Connossieur Cars
Connossieur	Fancy	Century Auto
Connossieur	Fancy	Connossieur Cars
Connossieur	Super	Century Auto
Connossieur	Super	Connossieur Cars
Ordinaire	Dull	Good Cars Inc.
Ordinaire	Dull	Ordinaire Dealer
Ordinaire	Dull	Orville Motors
Ordinaire	Boring	Good Cars Inc.
Ordinaire	Boring	Ordinaire Dealer
Ordinaire	Boring	Orville Motors

AUTOS

Figure 40: AUTOS

Each record gives the model, option and dealer. As shall be discovered, this relation above has the property that the insertion of one record requires the insertion of other records. For example, suppose the company starts to manufacture an additional option, called Plus 2, for the Ordinaire model. Because of the business rules, it must be made available to all the Ordinaire dealers. The records

Ordinaire	Plus 2	Good Cars Inc.
Ordinaire	Plus 2	Ordinaire Dealer
Ordinaire	Plus 2	Orville Motors

must be inserted into AUTOS. In like manner, if a new dealership for the Connoisseur model is acquired, three new records, one for each option, must be inserted.

What happens when a record is to be deleted? Suppose that the Connoisseur-Fancy is not selling well, and production is discontinued. At some point the dealers will no longer sell it, and the two records

Connoisseur	Fancy	Century Auto
Connoisseur	Fancy	Connoisseur Cars

must be deleted. This condition will surely lead to updating errors later on, resulting in inconsistent data in the file. As before, the solution is to decompose the relation so that two 4NF relations are formed,

- the first by taking a *projection* of AUTOS on the fields Model and Option to make the relation PRODUCTS
- the second by taking a *projection* of AUTOS on Model and Dealer to make the relation DEALERS (Figure 41).

Option	Model
Plus	Ordinaire
Standard	Ordinaire
Deluxe	Connossieur
Fancy	Connossieur
Super	Connossieur

PRODUCTS

Dealer	Model
Century Auto	Connossieur
Connossieur Cars	Connossieur
Good Cars Inc.	Ordinaire
Ordinaire Dealer	Ordinaire
Orville Motors	Ordinaire

DEALERS

Figure 41: PRODUCTS and DEALERS

In fact, one would probably want to decompose AUTOS into three files:

- one for models
- one for products
- one for dealers.

The resulting data structure diagram is shown below.

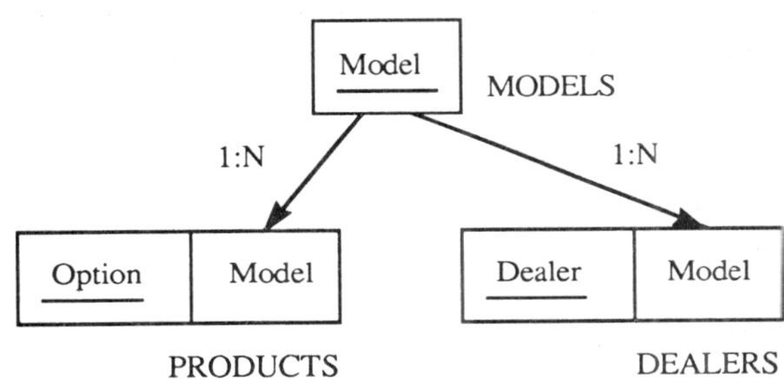

As you may have guessed by the title of this section, fourth normal-form files permit some undesirable dependencies. Although this is in fact true, in practice such examples are contrived and very rare. These dependencies are called join dependencies, and fifth normal-form relations have been defined to eliminate them. Generally, they are of a theoretical nature, and are beyond the scope of this text (Figure 42).

The procedure discussed in the preceding sections for deriving a logical database is summarized below.

1. Using the conceptual data model as the basis, explode any entities that are not relations into relational components (1NF files) and choose a key for each.

2. Eliminate undesirable dependencies (by decomposing the relations) as follows.

    a. Eliminate functional dependencies in composite key files where non-key fields are dependent on subkeys. (Convert to 2NF files.)

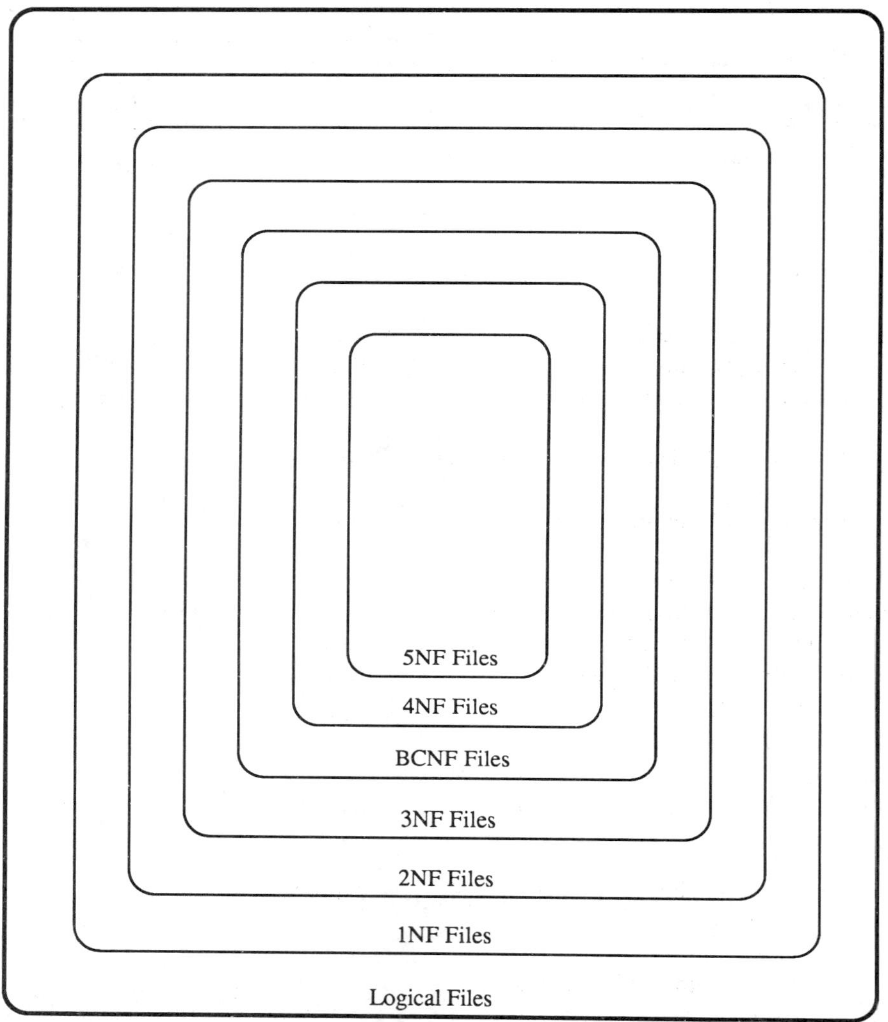

Figure 42: Logical files

b. Eliminate functional dependencies in which nonkey fields are dependent on other nonkey fields. (Convert to 3NF files.)

c. Eliminate functional dependencies in composite key relations where subkeys are dependent on nonkeys. (Convert to BCNF files.)

d. Eliminate multivalued dependencies. (Convert to 4NF files.)

3. Remove any remaining fields that are derivable by a process.

The database that results may be drawn in data structure diagram form (the schema).

4.5 Conclusions

This chapter has discussed the translation of the conceptual data model into a form suitable for automation: the logical or relational model. The objective of this effort is to produce a database having the following two attributes:

1. it satisfies the information needs of the user community, and

2. behaves well under automation.

(It has been the intention of the authors to provide some exposure to database fundamentals. Furthermore, we presume that in most environments, the corporate database would already be in place, and would not be derived as part of a particular development project. However, the analyst will use the database as part of analysis.)

4.6 Exercises

1. Why is it easier to change the logic of a process than the data that the process uses?

2. In the inventory file of Figure 20, suppose one supplier, Office Supply (OFC00) supplies several types of HB pencils, each with a different part number but having the same description. Is the composite field Description-Supplier still a candidate key for the file? Why?

3. Why in relational algebra is the second portion of the operator *project*, eliminating duplicates, necessary?

4. Why may normalization increase redundancy?

5. The following fields comprise the relation PERSONNEL:

- Empno - employee number
- Empname - employee name
- Supervisor - the employee's supervisor
- Supervisor-Level - the management level of the supervisor.

a. Determine all functional dependencies in the relation.

b. Which is (are) the undesirable dependency (dependencies)?

c. In which normal form is PERSONNEL?

d. Decompose PERSONNEL into a set of 4NF relations and make a diagram using data structure notation.

6. Using the E-R diagram of Figure 14 in Chapter 3 as a basis, derive a corresponding relational model.

7. A database used in the purchasing department requires information about suppliers, purchase orders and parts as follows:

- Supplier
    - supplier number
    - supplier name
    - addresses (several)
    - balance
    - terms
    - part (several)
    - part number
    - part description
    - part cost

- Purchase Order
  - order number
  - supplier number
  - supplier name
  - supplier address
  - order date
  - shipping detail
  - part number (several)
  - part description
  - quantity ordered
  - cost
  - total cost
- Part
  - part number
  - part description
  - quantity on hand
  - reorder level
  - reorder quantity

a. Create an E-R diagram for the above description.

b. Design a 4NF model for this database.

8. Use the relational algebra operators discussed in this chapter and the database developed in the previous question to find the following:

   a. Names and addresses of suppliers who supply part 127X4B.

   b. List of all parts and all suppliers for whom a purchase order has been issued on a given date.

   c. List of parts to order from supplier A (minimum cost).

9. The following functional dependencies are given between fields A, B, C, D, E, F.

   A,D--->B (A and D determine B)

   A--->E

   B--->C

   D--->F

   A,D--->C

   In what normal form are the following relations?

   a. R1(A,D,B,C)

   b. R2(A,C)

   c. R3(A,D,B,C,E)

   d. R4(A,D,C,F,B)

   e. R5(D,F)

10. The database for an investment firm consists of the following information:

   - For each broker B

     - his office number O

     - the list of investors I he looks after

- For each type of stock S
  - the dividend D paid by the stock
- For each investor I
  - the quantity Q of each type of stock owned by the investor.

Assume each investor deals with one broker. Give the functional dependencies and draw a set of 4NF files using data structure diagram notation.

11. A database to be used in the purchasing department of Quality Fabrics (QF) is to be built around the following information:

- Supplier
  - supplier name
  - address
  - balance
  - terms
  - pattern numbers (several - unique only to each supplier)
  - color numbers (several per pattern - unique only to each pattern)
  - type of fabric (woven, pile, wool, cotton, etc.)
  - price per yard
- Purchase Order
  - order number
  - supplier name
  - supplier address

- order date
- shipping detail
- supplier's pattern numbers (several)
- supplier's color numbers (several for each pattern)
- yards ordered
- cost
- total cost

- Inventory
  - QF pattern name
  - QF color name (several per pattern, unique to each pattern)
  - yards on hand
  - reorder level
  - reorder quantity

Design a 4NF model for this database and draw a data structure diagram, but first state all assumptions not implied (try to minimize) and weaknesses. Design your database based on your assumptions above.

12. Given the database shown below, which is for the personnel department of a company, what relational algebra operations would be necessary to answer the following:

    a. What departments are managed by vice-presidents?

    b. Who are the employees reporting to Mary Lamb?

    c. Who works in office 6002?

d. Assume that the attribute Date has the form yymmdd, and that you have available functions like min, max, avg, etc. During what period of time did Mary Lamb earn less than $20,000 (assuming no salary decreases)?

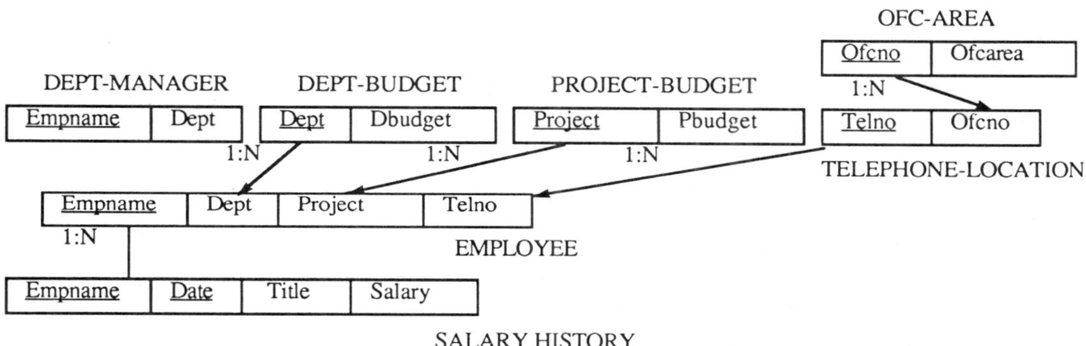

13. An engineering consulting firm with 37 consultants needs to track the progress of its many projects. It needs the information to properly bill clients for work done on projects, although they have an accounts receivable system to handle the actual billing process. This tracking application would also be used for management purposes to find out the following.

- How much billable vs. non-billable time is spent.
- How much overtime is being spent.
- How many projects each consultant is working on and how much time is spent on each project week by week.
- How many hours and dollars have been spent on each project this week, this month, this quarter, and to date.

- How actual hours and dollars compare with estimates made by each project manager.

- The status of the projects handled by each project manager.

Each consultant has an employee number and a specific billing rate. If the billing rate changes for an employee, a new employee number is assigned. A consultant may be working on many projects concurrently. For each project, a particular consultant is assigned to be project manager. Clients are billed at the standard billable rate for each employee even when the employee is working overtime. Clients are assigned a client number. There may be several projects from one client and more than one client may be involved with the same project. Project managers need to see summary reports for all their projects. Reports have to be sent to clients on the status of their project(s). Each week, consultants fill out time cards, recording their employee number, the week-ending date, and the number of regular and overtime hours worked on each project. Vacation time, sick time, travel time, and other miscellaneous overhead activities are charged to a series of project numbers from Z9000 through Z9999.

Create an E-R diagram for the consulting firm using the information listed below.

- Empno Employee number (4 position number)

- Empname Employee name (30 characters)

- Billrate Billing Rate for an employee ($30.00 to $100.00)

- Projno Project number (5 characters) an alpha character followed by 4 numbers e.g., A1234

- Projdesc Project descriptions (25 characters)

- Cnum Client number (5 characters) an alpha character followed by 4 numbers e.g., C1234

- Client Client name (30 characters)

- Claddr Client Address (30 characters)

- Reghrs Regular Hours worked as taken from the weekly time card (up to 99.9)

- Others Overtime hours worked as taken from the weekly time card (up to 99.9)

- Tothrs Total hours worked as taken from the weekly time card (up to 99.9)

- Projmgr Project manager (employee number)

- Esthrs Estimated hours to complete the project from start to finish (up to 99999)

- Estdol Estimated dollars to complete the project from start to finish (up to 99999)

- Pctcom Percent complete for the project (0 to 100)

- Estcom Estimated completion date for the project

- Wdate Week-ending date

- Projloc Project location (30 characters)

14. Translate the conceptual model developed in Exercise 3 of Chapter 3 to the relational model.

15. Design a database for storing all the information necessary for exercises 4a and 4b of Chapter 3.

16. Translate the conceptual model developed in Exercise 10 of Chapter 3 to the relational model.

17. Discuss the possible different uses that E-R diagrams and data structure diagrams may have in software development.

4.7 Bibliography

1. Albano, A.V. de Antonellis, and di Leva, A. *Computer-Aided Database Design*. Amsterdam, The Netherlands. North-Holland, 1985.

2. Armitage, Howard M. and McCarthy, William E. *Decision Support Using Entity-Relationship Modeling. Journal of Accounting and EDP*. Fall, 1987.

3. Bachman, C.W. *Data Structure Diagrams*. Data Base Journal of ACM SIGBDP 1, No. 2, 4-10. 1969.

4. Bradley, James. *An Introduction to Data Base Management in Business, Second Edition*. New York, New York. Holt, Rinehart and Winston, 1987.

5. Chen, Peter. *The Entity-Relationship Approach to Logical Data Base Design*. Q.E.D. Monograph Series, No. 6. Wellesley, Massachusetts. Q.E.D. Information Sciences, 1977.

6. Codd, E.F. *Recent Investigations into Relational Data Base Systems*. Proc. IFIP Congress, 1974.

7. Codd, E.F. *Relational Completeness of Data Base Sublanguages*. Data Base Systems, Courant Computer Science Symposia Series, Volume 6. Englewood Cliffs, New Jersey. Prentice-Hall, 1972.

8. Dampney, C.N.G. et al. *Data (Information) Analysis of Business Systems - Informal Notes*. Waterloo, Ontario. University of Waterloo, 1986.

9. Date, C. J. *An Introduction to Database Systems, Volume 1, Fourth Edition*. Reading, Massachusetts. Addison-Wesley, 1986.

10. Davis, G.B. *Management Information Systems - Conceptual Foundations, Structure and Development*. New York, New York. McGraw-Hill, 1983.

11. Fagin, R. *Multivalued Dependencies and a New Normal Form for Relational Databases*. ACM TODS 2, No. 3, September, 1977.

12. Furtado, A.L. and Neuhold, E.J. *Formal Techniques for Data Base Design*. Berlin, West Germany. Springer-Verlag, 1986.

13. Howe, D.R. *Data Analysis for Data Base Design*. London, U.K. Edward Arnold, 1983.

14. Korth, Henry F. and Silberschatz, Abraham. *Database System Concepts*. New York, New York. McGraw-Hill, 1986.

15. Martin, James. *Managing the Data Base Environment*. Englewood Cliffs, New Jersey. Prentice-Hall, 1983.

16. Norris, Fletcher R. *Discrete Structures: an Introduction to Mathematics for Computer Science*. Englewood Cliffs, New Jersey. Prentice-Hall, 1985.

17. Parkinson, Richard C. *Data Analysis: the Key to Data Base Design*. Wellesley, Massachussetts. QED Information Sciences, 1985.

18. Tsichritzis, Dionysios and Lochovsky, Frederick H. *Data Models*. Englewood Cliffs, New Jersey. Prentice-Hall, 1982.

Chapter 5

DATA MANAGEMENT

The previous two chapters have dealt with the global corporate data resource and its modeling. The aim of such modeling is to permit effective utilization of an important resource. The global data model produced provides a unified view of data and at once lends itself to automation and serves the user community.

Effective data utilization requires proper management. The rise in importance of data together with advances in database technology have caused management of the data resource, apart from the systems using the data, to become a goal for many organizations. As management's awareness of data has increased, tools and methods to effect data management have been developed. The current "database era" has led to new methods of managing data, for designing and developing software (discussed in Part 3), to new problems in data security, privacy, redundancy, integrity/ambiguity and tracking, and to new functions (data administration) [1]. From a corporate view, some objectives of data management may include

- easier access to data for problem solving and decision making
- elimination or control of data redundancy
- standardization of data, so that data may be shared and interchanged among various functional areas
- data security
- data integrity [13].

To treat data as a resource implies that there are characteristics of data which are common to other resources, such as capital and people. These resources have sources of supply, destinations for their use and levels of availability. In order to manage the data resource effectively, one must learn as much as possible (obtain as much data as possible) about the organization's data. Centralized co-ordination and control is required. A *data administrator* or *information officer*, analogous to the financial officer or personnel officer of an enterprise, is responsible for the data resource. The data administration function is responsible for developing and administering policies, procedures, plans and practices for the definition, organization, protection and usage of data within an enterprise. The scope of the data administration function includes *all* corporate data, and not simply those data that are part of an automated database management system (DBMS).

How are data resources managed? They are managed through administration and control, just as any other resource is managed. In order to effect data resource management, the enterprise must know as much about the data as possible: who is responsible for the data, where does the data originate and reside, and who uses the data? Such data about data are sometimes referred to as *metadata*. Note that the emphasis in data management is on the information about the data, and not on the data per se.

5.1 Data Dictionaries/Data Directories

A *data dictionary* (or *data directory*) refers to the primary tool used in support of data management. It can be viewed as a database (or central reference) describing all the details of the data of an enterprise, and might include items such as:

- what the data are and mean
- their attributes and structure
- their relationships
- where they are located
- who is responsible for the data
- where they are used
- how the data can be accessed.

A data dictionary or data directory *system* (DDS) is a system for managing the data dictionary. Data dictionaries and data dictionary systems may be manual or automated, and automated ones may be dependent upon a DBMS, or may stand alone. They provide for the logical centralization of the information about an organization's data resources [7]. A DDS, then, is a centralized reference library, which in practice may vary from a simple, manual, 3 x 5 card system to an automated, dynamic DDS which is accessed at execution time, together with the manual and/or automated procedures prescribed by the enterprise. A DDS is a mechanism for recording and processing information about the structure and usage of all the working data of an organization, and for the management, control and handling of the data.

A data dictionary system should be able to store and manage definitions for

- an organization's data, regardless of where that data are located – in branch office filing cabinets, on documents or on computer storage files
- an organization's procedures, be they clerical, managerial, or automated
- an organization's systems, both manual and/or automated
- the users of systems and data – divisions, departments and people [12].

A random selection of commercially available data dictionary systems is listed below.

PRODUCT	VENDOR
Datamanager	MSP, Inc.
Encyclopedia	KnowledgeWare, Inc.
DB/DC Data Dictionary	IBM
IDD	Cullinet
Data Catalogue 2	TSI International
ADABAS Dictionary	Software AG
PRIDE IRM	M. Bryce & Assoc.
UCC-10	University Comp. Co.
Dictionary/3000	Hewlitt Packard
DDS-1100	Sperry-UNIVAC
In-Line Directory	Cincom Systems

Historically, data dictionaries have been viewed from the systems development perspective. However, through technological advances and the rise in importance of data, they are now viewed as being a powerful management tool.

Because a data dictionary system provides information about where data are used, it should (ideally) identify to software developers (and others) the impacts of changes to particular data. DDS's standardize data, and allow users to communicate with each other using common terms and definitions.

On a less pervasive level, a data dictionary is a tool for management of the data processing (DP) function. If all projects must adhere to the definitions and conventions laid out in the corporate data dictionary, then the data of various projects are standardized. This is especially beneficial for corporations in which distributed processing occurs, because if each local DP installation develops all applications according to the data dictionary, then none can develop incompatible applications and all parts of the organization can share data.

In many corporations today, data management is a goal, a goal not yet reached. Symptoms indicating a lack of data management include

- an inability to enforce data processing standards
- data ambiguity
- data redundancy
- an inability to determine the effects of a data change
- an inability to identify the users of data and the processes in which the data are used [5].

How has such a situation been permitted to develop? Expediency and the lack of experience are the main culprits. In order to get software systems implemented as quickly as possible, systems developers (including analysts) have obtained the required data from the least expensive or most expedient source. This probably meant obtaining some data from existing files and massaging it to suit the particular system being developed. Because the systems development function is such a new one in companies, and because since its inception it has been under pressure to deliver systems as quickly as possible, the path of least resistance has been the one followed. Today, with the natural maturity that comes with experience, organizations are realizing in retrospect that this approach has resulted in the current "data mess" found in many corporations: the same data exist in different applications in different formats and having different names, with the potential/reality of inconsistency and incompatibility. Then in order to get the corporate data into shape, data administration has four main functions.

- standardize data definitions
- control shared data definitions
- manage distributed data
- oversee security functions [12].

We conclude this brief overview of data managment and data dictionary systems with an example of how the entries for a particular entity might appear.

5.2 Example of Data Dictionary System Usage

Data dictionaries may be manual or automated, but the principles are the same. Each entry in a manual data dictionary may be on an 8 1/2 x 11 sheet of paper, filed alphabetically. In KnowledgeWare's Encyclopedia, which is part of KnowledgeWare's Information Engineering Workbench (IEW), as data are entered into the IEW, the Encyclopedia stores such entries, as well as the rules governing their use. "Where used" reports can be produced, showing the systems that use particular data. As well, the Encyclopedia indicates the impacts of changes to particular data [9]. A discussion based on a generic manual data dictionary follows.

One entity type of major interest to many enterprises is that of ORDERs. In general, much money and effort is devoted to marketing which, it is hoped, will result in increased market share (more orders from customers). Suppose we wish to record in a data dictionary the details of this entity. A good starting point is to examine a sample sales order form (Figure 43). The order is made up of information pertaining to the order as a whole (header), information about the individual items ordered (body) and summary information following the body (footer).

The structure can be initially represented as

```
ORDER

    Order-Header   }   components
    Order-Body     }   of
    Order-Footer   }   order
```

ORDER: 12573 SALES ORDER DATE: ____________

FIRST CITY COMPUTERS

Bill to: Ship to (if different):

Terms: COD ☐ Cash ☐ 2% 10, net 30 ☐ 60 days ☐

Quantity	Description	Unit Price	Total

Subtotal: ____________

Tax: ____________

Ordered by: ____________ Total: ____________

Figure 43: Sample sales order

The details of the data structure Order-Header can be specified as follows:

Order-Header
  Order-Number
  Order-Date
  Bill-To-Address
  [Ship-To-Address]   [ ] indicates optional structure

Terms
  <COD>
  <Cash>
  <2% 10, net 30>   <> indicates alternatives,
  <60 days>   one of which is applicable

Because the structure of Order-Body consists of repeating occurrences of each Order-Line, it is possible to describe the structure of Order-Body as

Order-Body
  Order-Line*   * indicates possible repetitions
    Qty-Ordered
    Item-Description
    Unit-Price
    Line-Total

There are eleven lines within the body of the order, so Order-Line may occur one to eleven times. This is indicated by

Order-Line*(1-11).

The composition of each data element comprising the data structures Order-Header and Order-Line (Order-Number, Order-Date, Bill-To-Address, etc.) is obvious. Line-Total is defined by the calculation

Line-Total

Quantity-Ordered x Unit-Price

Now the structure of ORDER has been described completely, and has been specified from the top level down to the lowest or data-element level.

The data dictionary entry for **ORDER** appears in Figure 44. Each component of **ORDER** must also have its own entry, and most of these are illustrated in Figures 45 through 52 at the end of the chapter.

ENTITY: ORDER

ALIAS: SALES ORDER, CUSTOMER ORDER

DESCRIPTION: Source document written by sales personnel containing details of each items ordered and signed by customer

COMPOSITION:

Order-Header
Order-Body
Order-Footer

WHERE USED:

Order Processing System (OPSYS)

processes 1.1.1 (VERIFY CUSTOMER ACCOUNT)
1.1.2 (UPDATE INVENTORY)

Sales Analysis System (SALANAL)

processes 5.2 (CALCULATE SALES COMMISSIONS)
7.2.1 (CUSTOMER PURCHASE HISTORY)

OWNER: marketing division

HOW ACCESSED: through accounting database (ACCTDB)

Figure 44: Data dictionary entry for ORDER

Notice that SALES-ORDER and CUSTOMER-ORDER are aliases (other names) for ORDER. There may be several aliases for any given data structure or element, and conversely, a term may have several meanings, depending upon the context. Clearly, this is one reason why the data dictionary exists. As well, a description of the data structure is included which provides general information, and the composition or structure is also specified. Other aspects of this data dictionary entry include

- where used - a listing of the systems, both manual and automated, using the entity ORDER (this listing facilitates the tracking of the impact of a data change)
- owner - who is responsible for this entity
- how accessed - how one can access this entity.

A complete DDS contains all E-R models, relational models and their components and descriptions of all processes in all systems. For software developers working on a particular project, the data dictionary is a central reference for data definitions. As the tools for systems analysis are discussed in Chapters 7 and 8, the importance, usefulness and contents of the data dictionary will become more apparent.

5.3 Exercises

1. How can a DDS assist in estimating costs of planned modifications to systems?

2. Auditors, both company auditors and government auditors, are concerned with the control and accuracy of data. What role can a DDS play in the audit function?

3. The purpose of a data dictionary system is to solve the problems caused by variations in the enterprise data that are interchanged and shared, and to improve control and communications throughout the corporation. To achieve this purpose, a DDS must have a sound basic structure. Prior to building a DDS, what major steps do you feel should be taken to ensure a proper foundation.

4. What role does a DDS play for programmers during the programming phase of software development?

5. Search out in the library, or in companies familiar to you, two commercial data dictionary systems. What are the major features of each? How do the products compare, and which one do you favor over the other? Justify your choice.

6. Design a form that could be used for making data entries into a manual data dictionary system.

7. List some examples of data ambiguity in your organization.

8. Name three entity types or data structures having more than four aliases.

9. Compose data dictionary entries for all the entity types, relationships and attributes for exercise 2 in Chapter 3, as well as one entry for the E-R model.

10. Compose data dictionary entries for all the entity types, relationships, and business functions described in exercise 4 in Chapter 3.

11. Compose data dictionary entries for the database described in exercise 7 of Chapter 4.

12. Compose all the data dictionary entries for exercise 13 of Chapter 4.

13. Suppose you have been hired to be the data administrator for a company. Because funds are limited, you decide to establish a manual data dictionary system. The company has automated some of its accounting procedures: order processing, inventory control, accounts receivable and financial reporting (financial statements). How would you organize a data dictionary system that would describe the corporate data, the processes and systems and other information necessary for implementing an adequate DDS? Compose sample entries to illustrate your method of organization.

14. How can a DDS help users?

15. Implicit in this chapter is the view that the systems department should focus on data and information management, in addition to getting systems developed as quickly as possible. Discuss the pros and cons of this view.

16. Do you believe that all enterprise data should be under the control of the data administration function, or only data that are shared among different organizational units? Substantiate your view with sample scenarios.

17. Visit a comporation or organization having a data administration function and find out

    a. what the data administration function perceives its role within the organization to be

    b. what types of data problems are currently being worked on

    c. the types of data the function has responsibility for

    d. what kind of DDS the organization has.

5.4 Bibliography

1. Braithwaite, Ken S. *Data Administration.* New York, New York. John Wiley & Sons, 1985.

2. DeMarco, Tom. *Structured Analysis and System Specification.* Englewood Cliffs, New Jersey. Prentice-Hall, 1979.

3. Gane, Chris and Sarson, Trish. *Structured Systems Analysis: Tools and Techniques.* Englewood Cliffs, New Jersey. Prentice-Hall, 1979.

4. Gorman, Michael M. *Managing Database: Four Critical Factors.* Wellesley, Massachusetts. QED Information Sciences, 1984.

5. *A Guide on the Role of the Data Dictionary in the Information Management Process.* Ottawa, Ontario. Government of Canada, 1985.

6. Lefkovits, Henry C. *Data Dictionary Systems.* Wellesley, Massachusetts. Q.E.D. Information Sciences, 1977.

7. Leong-Hong, Belkis W. and Plagman, Bernard K. *Data Dictionary/Directory Systems.* New York, New York. John Wiley & Sons, 1982.

8. Lomax, J.D. *Data Dictionary Systems.* Manchester, England. NCC Publications, 1977.

9. Martin, James. *Information Engineering.* Carnforth, England. Savant Research Studies, 1986.

10. Mayne, A. *Data Dictionary Systems: A Technical Review*. Manchester, England. NCC Publications, 1984.

11. Ross, Ronald G. *Data Dictionaries and Data Administration*. New York, New York. AMACOM, 1981.

12. Sprague, Ralph and McNurlin, Barbara. *Information Systems Management in Practice*. Englewood Cliffs, New Jersey. Prentice-Hall, 1986.

13. Van Duyn, J. *Developing A Data Dictionary System*. Englewood Cliffs, New Jersey. Prentice-Hall, 1982.

DATA STRUCTURE: Order-Header

ALIAS:

DESCRIPTION: Customer information portion of order

COMPOSITION:

Order-Number
Order-Date
Bill-To-Address
[Ship-To-Address]

Figure 45: Data dictionary entry for Order-Header

DATA ELEMENT:	Order-Number
ALIAS:	ON, customer order number
DESCRIPTION:	Preprinted number on order forms

VALUES, COMPOSITION or CALCULATION:

10000 - 99999

Figure 46: Data dictionary entry for Order-Number

DATA ELEMENT: Ship-To-Address

ALIAS: Ship-to, shipping location

DESCRIPTION: Used in customer orders when shipping address differs from billing address

VALUES, COMPOSITION or CALCULATION:

1 name line
1 - 3 address lines

NOTES:

Figure 47: Data dictionary entry for Ship-To-Address

DATA STRUCTURE: Order-Body

ALIAS:

DESCRIPTION: Details of items ordered by customers

VALUES, COMPOSITION or CALCULATION:

Order-Line *(1-11)

Figure 48: Data dictionary entry for Order-Body

DATA STRUCTURE: Order-Line

ALIAS:

DESCRIPTION: Description of individual items and quantities

VALUES, COMPOSITION or CALCULATION:

Quantity-Ordered
Item-Description
Unit-Price
Line-Total

Figure 49: Data dictionary entry for Order-Line

DATA STRUCTURE: Line-Total

ALIAS:

DESCRIPTION: Value of ordered quantity for individual items

VALUES, COMPOSITION or CALCULATION:

Quantity-Ordered x Unit-Price

Figure 50: Data dictionary entry for Line-Total

DATA STRUCTURE: Order-Footer

ALIAS:

DESCRIPTION: Portion of order pertaining to total amount and authorization

COMPOSITION:

Customer-Authorization
Order-Subtotal
[Order-Tax]
Order-Total

Figure 51: Data dictionary entry for Order-Footer

DATA STRUCTURE: Order-Subtotal

ALIAS:

DESCRIPTION: Value of goods on an order

VALUES, COMPOSITION or CALCULATION:

sum of all Line-Totals in a given ORDER

Figure 52: Data dictionary entry for Order-Subtotal

Part 2

Elements of Systems Analysis

Part 2 provides an overview of the traditional approach to software development. Recall that the six phases of the traditional life cycle are:

1. Requirements Determination
2. Systems Analysis
3. Design
4. Programming
5. Testing
6. Implementation.

The life cycle is shown in Figure 53 as a data flow diagram.

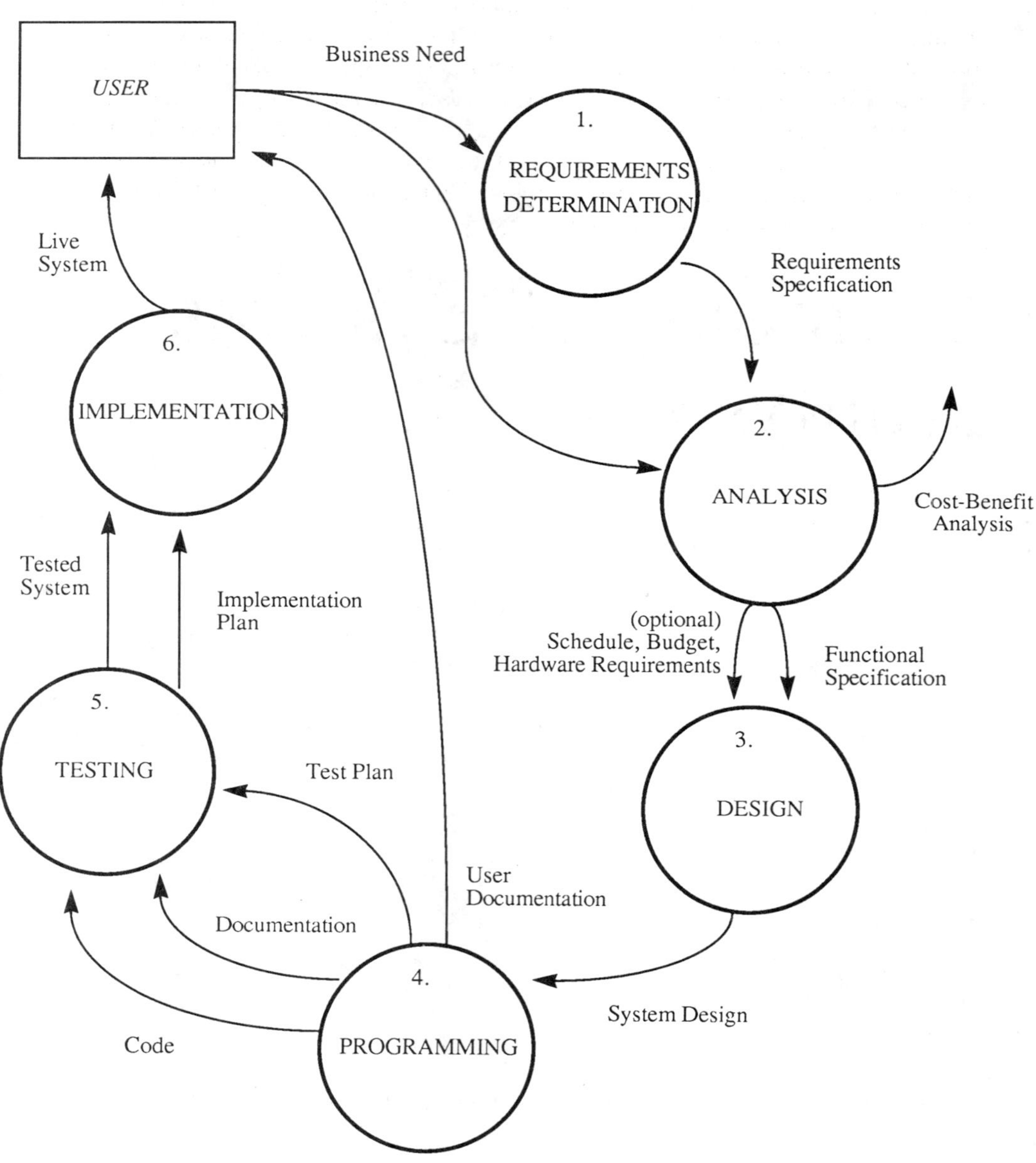

Figure 53: Data flow diagram of the traditional life cycle

The first two phases in the development cycle of a prospective system (requirements determination and systems analysis) are concerned with a high-level determination and evaluation of business requirements, followed by a detailed analysis and conceptual design of a system being created to meet those requirements. The remaining phases, from design to implementation, will not be covered in this text because they are of an entirely different nature, being much more straightforward and well defined. Part 2 begins with an overview of the origin and evaluation of requests for systems.

Chapter 6

THE PROCESS OF ANALYSIS

6.1 Before Analysis: Requirements Determination

The requirements determination phase of a system's life cycle, which is closely coupled with the analysis phase, is important enough to cover as a separate topic. The reason for this is that a systems analyst must understand the user's needs sufficiently to transform business requirements into systems requirements. It is essential that the analyst understand the origin and objectives of the project, and why it is being requested. As the analysis phase progresses, the original requirements will probably be refined and perhaps modified. The requirements phase, in fact, is never really complete, for even after a system is functioning, changing requirements will most certainly call for enhancements and modifications.

6.1.1 Needs Identification

Before any involvement with systems people occurs, there is always background work required of prospective users. This work involves the identification and documentation of the often-mentioned business needs targeted for computer support. Some examples of typical business problems that appear in need of automated aids follow. These examples are from the authors' own experiences and are obviously simplified.

1. A medium-size manufacturing corporation has five branch offices throughout the country which fill orders for equipment and send copies of these orders to head office. Head office in turn prepares invoices which are mailed to customers for payment. The total turnaround time from the

branch receiving an order to head office receiving payment for the invoice is 72 days. This delay is judged unacceptable by the company's owner because of the potential for losing bank interest and the difficulty in obtaining up-to-date information on the status of company sales and inventory.

The problem appears to be that most of the order processing procedure is manual, relies on mail service and is redundant. The owner sees a need for automation of some sort to solve or alleviate this problem.

2. The president and several senior vice-presidents of a major corporation believe that they are not receiving daily and weekly information as quickly as possible. They receive reports which have been prepared manually by the extraction of data from batch computer systems. Most of the data is in table form, but it is produced as hand-drawn charts. These managers feel that their information is not as timely as it could be and that much information exists that they do not receive at all. They believe the data is somewhere in the company computers.

   A task force of managers is formed to study the situation, to detail what information the senior managers require, to briefly investigate what other corporations are doing in terms of executive automation and to plan a strategy for developing an information system for all senior executives. Clearly, this is purely a data processing application, but before involving systems people, the managers want to examine the problem from the viewpoint of their executives' requirements. Some sample information requirements include:

   - The monthly and yearly corporate budgets expressed in terms of year-to-date expenditures, budgeted expenditures and outlooks are required. Because the data exist at a detailed level in this large corporation, it is necessary to aggregate the data to provide an overall picture of the state of the company, and how the state compares to budgets and outlooks.

   - Weekly summaries of employment status levels in all divisions – resignations, retirements, full-time and part-time employment – are required. These summaries are to be compared to budgeted levels.

   - Figures for the previous day's sales are desired by 7:30 a.m. each morning.

There are many other requirements too numerous to list here. After documenting all information requirements, as well as performance characteristics (fast response, graphics), data processing was contacted and a project team was formed to analyze and ascertain feasibility of the project.

3. The management of a large service corporation is upset at how long the response to a customer complaint takes (6 weeks). The process in place is completely manual: when a customer complaint is received, copies of the complaint are sent to appropriate departments for investigation. After several weeks, the contacted departments send back their findings and a letter is prepared by a customer relations person. The manager of customer relations envisions a system based on word processors with sample replies so that no new letter composition would be required, only minor changes to a standardized letter. The manager also thought of eliminating any contact with other departments except in special cases, the rationale being that a fast response is much better than a detailed excuse.

   In this case, the manager not only had a problem, but also a solution in mind. However, corporate procedures required that data processing be contacted.

4. The marketing group of a major transportation company wishes to create better products and marketing plans by developing a decision-support system based on new concepts not previously tried in the industry. The new concept utilizes historical market figures together with economic data to forecast regional product demand. In addition, simulation models and information systems will provide the capability of evaluating new products and scenarios. A team of marketing analysts led by a senior manager works out a high-level description of an idealized way of working with these concepts.

5. The owner of a small roofing company whose procedures are completely manual perceives that a word processor or other automation aids (i.e., software for job costing) would help run the business more effectively. The owner is responsible for the office function of the business, with the help of laborers and believes that automation would be more beneficial than hiring an additional person to work in the office. The owner believes that the needs of the business, although vaguely formulated, could be met by automation.

The above examples are based on actual experiences in modern companies. One can see that often much thinking and effort is expended before systems people beome involved. Often the user or customer has a solution in mind, and it is important for the systems analyst to envision a bigger (corporate) picture of a problem rather than being too influenced by the user's solution. At the same time, the analyst must realize that the user knows the business and its needs better than anyone else. The analyst must therefore, listen and learn. Is is always a problem trying to create practical, quickly delivered solutions which are at the same time well thought out.

Once the user has identified and documented all requirements, including benefits, it is time to present the case to senior management. After gaining their support, the user proceeds to data processing to initiate a project. At this stage, the benefits of automation are of prime importance to user and management alike. Technical feasibility and costs are secondary, mainly because at this point they are usually unknown. For many applications, benefits are tangible because they can be evaluated and quantified fairly easily. A typical example is a company in which manual procedures are replaced by automation.

There are also many cases where projects are approved on the basis of so-called *intangible* benefits – those impossible to quantify in dollars. These are often related to a company remaining competitive. For example a bank without automatic teller machines (ATM's) or an airline without advanced seat selection will surely lose market share to competitors offering such services. Another example of intangible benefits is supplying better information faster to management, making them more effective. A dollar figure for such benefits may be meaningless. In any case, once the tangible and intangible benefits have been identified, the next step involves presenting requirements to data processing personnel in terms of a proposal, and forming a project team or task force to evaluate the proposal.

6.1.2 Understanding and Evaluating Requirements

During requirements determination, the user's identified requirements must be appraised in terms of economic feasibility and potential for automation. This phase requires that those data processing personnel involved gain a certain degree of understanding of the user's business requirements and stated requirements. The mode of operation is usually a small project team or task force, with the systems analyst often the only data processing person assigned. Sometimes the data person assigned is not a systems analyst, but a manager or superior of the analyst. The time spent and the number of people involved in requirements determination depends on the priority, complexity and size of the application; normally this phase is, at most, several weeks long. The end result will be a document (requirements specification document) outlining the present business environment, the need for change, objectives to be met, expected benefits, automation potential, feasibility and rough cost estimate.

The means by which the team of users and systems people gains the required knowledge varies according to both the magnitude and type of application. A minor, straightforward problem may require only preliminary discussion with, or a presentation by, the users. Most often, though, problems are complex and require a more organized approach, often according to a work plan. At this stage, it is very important that excessive effort not be expended exploring the processes in great detail and gathering more information than necessary. The goal here is to assess the situation faced by the user and to gain only as much knowledge as is required to make a decision whether or not to proceed with an analysis. That is, does an opportunity for automation exist whereby current processes are eliminated, changed or directly automated? Or are there totally new ways of doing business or of managing the current business that are not possible without a new system? The decision is made on the basis of benefits, feasibility and costs.

The most common method followed in determining requirements is for the analyst to interview selected individuals, starting with management and continuing down to an appropriate staff level. It is best that the analyst have a list of questions prepared in order to be consistent and comprehensive. Normally two interviewers are adequate, as more tend to make the respondent ill at ease. One interviewer should be a user representative from the project team and the other a systems person from the team. Although prepared questions are available as a guide, much learning occurs when discussion leads to peripheral topics. Remember that the goal of this exercise is to gain an understanding of the situation; it is not a problem-solving exercise. Very often users will suggest system solutions right at the beginning of an interview, such as "what we need here is a P.C. with spreadsheet software and a link to our mainframe... ."

It is difficult not to be influenced at this stage by what appear to be valid suggestions to an, as yet, partially defined problem. But it is important to keep in mind the corporate viewpoint in order to first gain a good understanding of the total business problem. Many human dimensions are linked with the problem too, not just automation. Furthermore, one may often find situations where different, and apparently conflicting, responses arise from persons of different levels. For example managers usually perceive operations differently from their employees. Employees, on the other hand, will often know more than their managers about how the business operates on a day-to-day basis.

As an example illustrating this last statement, consider the manufacturing company mentioned previously. The office staff in the western branch of the company always calculated the amount owing for each order shipped to one of its customers. This procedure evolved into an informal but accurate method for calculating the amount owed by each customer. When customers sent in payments to the branch, the payment was deducted from the amount owing before the payment was forwarded to the company's corporate office. The company president, when interviewed about the needs of the company, stated that the branches had no current information readily available to them concerning the status of customer accounts.

Another form of gathering information for requirements determination is the questionnaire, especially useful where large numbers of people are involved. It is critical, however, that the author of a questionnaire be knowledgeable about the area being investigated. Questions relate to the effectiveness of existing processes, exceptions that occur, existing problems and so on. Both multiple choice and open-ended questionnaires may be used. Although they are easy to administer and the results are readily compiled, the tradeoff with questionnaires is a loss of flexibility and adaptability.

At this stage, the role of the systems personnel is to contribute their knowledge, experience and judgement regarding the feasibility of automation, and to ask the right questions in order to obtain the right information.

6.1.3 Documenting and Communicating Results

Once the business area is understood and requirements have been analyzed, and presuming there is sufficient potential for a system to be developed, the next step is to document the findings and communicate them to appropriate levels of management. A written document, the requirements specification, contains background information about the business environment, the need for change, an analysis of requirements, possibly a list of alternatives, and a specific recommendation. Estimates of costs, benefits and opportunity loss (i.e. the cost of taking no action) are included, even though they are very rough. In addition, intangible benefits (they may, in fact, be the only benefits) are identified, and a clear explanation of the impact of the proposed approach is given. Other items to possibly include with the requirements specification are

- assumptions
- risks
- estimated development time
- a preliminary workplan for the analysis phase.

Sometimes it is not feasible to proceed directly with systems analysis because new methods of work or new processes must be developed prior to its commencement. For example, the electronics manufacturer considered previously may have also determined that not all of the problems concerned their method of filling orders. It was observed that part of the reason for the current situation was that orders were physically filled from several different locations: there were several warehouses and redundant inventories. Before designing a new system it was essential that a major study be undertaken to create the most efficient means of filling orders. This study included deciding between a centralized single-inventory warehouse and corresponding computer system, or continuing with the multiple-warehouse, multiple-inventory set-up. Further considerations included locations, optimal stock levels and balancing costs against customer satisfaction. Such problems are not trivial, and they must be solved before a system is designed or else the result may be the automation of a costly mistake.

Thus the requirements phase should address a business area from the viewpoint of creating the best possible process, one that will be supported by a system that attains the best results from the process.

6.1.4 Origins of Projects and Current Trends

From the examples discussed in this chapter, it should be clear that there is considerable diversity in systems projects with respect to both size and complexity of the application. Projects may originate from any area within an organization, and as the business environment, both internal to and external to the company, changes, so do trends in development projects. This section concludes with some remarks concerning trends in projects undergoing automation (or reautomation).

The applications areas currently being developed are different from what they have been historically. Traditional operational support functions, such as financial reporting, inventory control, payroll and so on, have already undergone automation. Higher-level applications areas whose products are information are now being targeted for automation. Decision support systems (DSS) are often used to describe such systems. The classical management triangle, first described by Robert Anthony in 1965, is useful for describing systems (Figure 54).

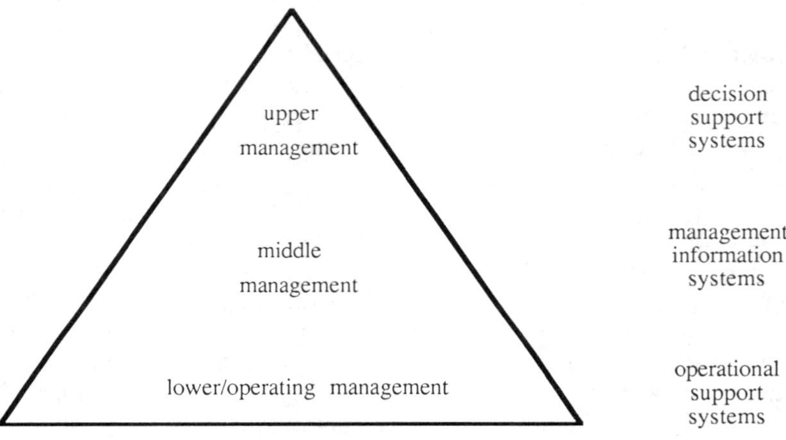

Figure 54: Management triangle

The kinds of systems that can be automated are shown beside the triangle. Regular business operations such as those listed above require structured, completely predictable data as input. These operations usually involve automation of some previously manual processes, and they often summarize some aspect of the company's operations. Most often they are transaction based.

The other types of systems, those that support middle and upper management, respectively, are management information systems (MIS) and DSS; both MIS and DSS deal with information. Their purpose is not to perform a regular business process, but rather to provide management with information that has previously been unobtainable (or not easily acquired) and that will help management perform their job. Management information systems and decision support systems assist managment in making, implementing, monitoring and controlling business decisions.

As indicated by the triangle, operational support systems directly impact lower management. For example, in an accounts receivable system, there are probably clerks who apply payments to customer accounts, produce invoices and prepare customer statements. There may be an accounts receivable manager who approves/rejects credit for questionable customers and deals generally with policy matters concerning receivables, but who is not normally a direct user of the system. Summary information regarding accounts receivable (A/R) is probably provided to middle and upper management as part of the financial accounting and reporting function that takes place within the corporation, but middle and upper management are not directly impacted by the day-to-day operation of such a system.

On the other hand, information systems may directly affect both middle and upper management. In the recent literature, a distinction between these two kinds of information systems has arisen, as is illustrated by the management triangle. The distinction between decision support systems and management information systems is generally that, although both provide management with information that aids them in doing their job, DSS tend to be aimed at higher management, and MIS at middle management. Both use information that is less structured than that used in operational support systems, and the information required by DSS is even less structured than that required by MIS.

Decision support systems usually involve mathematical modeling, forecasting and simulation (refer to examples discussed earlier in the chapter). In developing such systems, it is *vital* that the analyst acquire understanding of the business area and of the information needs of the potential users. Adequate technical skills are obviously not the only requirement of the analyst in developing decision support systems. The interpersonal skills of the developer are essential to the success of any information-centered project. The ability to listen, to be perceptive enough to ask the right questions, to maneuvre through difficult situations and to be diplomatic when necessary, all contribute to understanding what the user wants and ultimately, to delivering a quality product. The ability to communicate effectively, in addition to adequate technical skills, is essential for systems analysis.

6.2 The Phases of Analysis

The previous section examined the first phase of software development, requirements determination. During this phase, the present business environment is documented and the need for change is outlined. The analyst or analysts who have been assigned to the second phase of the software project, analysis, may or may not have been involved during requirements determination. If they were involved, then a good start toward understanding the user's current situation has already been made. If an analyst has not previously been involved, then the documentation produced during the requirements phase can provide a starting point toward understanding the current system.

The analysis phase consists of six steps, followed more or less sequentially. These steps are:

1. Model the user's current physical environment

2. Abstract the underlying logic to produce a logical model

3. Derive a model of the new system

4. Develop automation strategies

5. Select a particular strategy

6. Package the selected system.

A data flow diagram (DFD) of analysis (Figure 55) shows how the steps relate to each other.

6.2.1 The Current System

The quality of the software ultimately produced depends on many factors. The most important factor is the quality of the analysis, and this, in turn, depends on the analyst's understanding of the current business environment. If the understanding is lacking or incomplete, then any proposed solution cannot possibly work. The analyst must examine all documents used in the current system and the information the system generates, and must interview the people who use the system and those who supply it with information. It is likely that the analyst will be talking with people who have different perceptions of what the current system does, its shortcomings and the relative importance or usefulness of the information (or lack of information) obtained. The analyst must sift through perceptions, documents and information to derive an accurate assessment of the current environment.

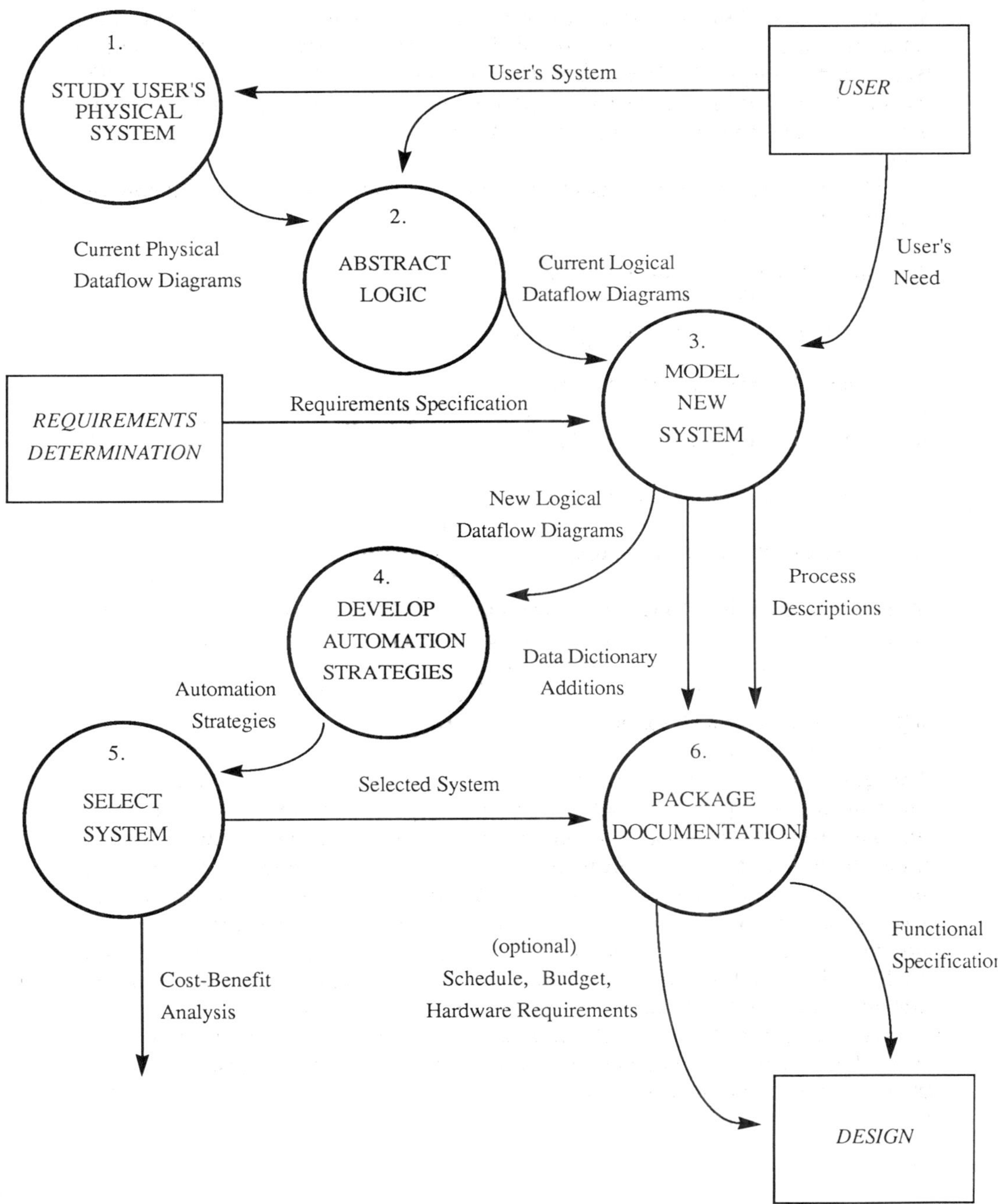

Figure 55: Data flow diagram of the steps of analysis

For example, suppose the system undergoing analysis is an automated order entry system. One of the reports produced by the current system is a monthly sales analysis by customer for each salesperson, with a breakdown of taxable and nontaxable sales. If the analyst were to ask a member of the salesforce the purpose of the report, the salesperson might answer that it facilitates a quick calculation of the commission earned for the month, and provides as assessment of customer activity. For the accountant or controller, however, it may serve as a basis for remitting federal sales tax. If management also receive copies of the report, it is probably useful to them for assessing the ongoing performance of the salesforce. Each group has a different view of the report's purpose, and all aspects contribute to the total picture.

In modeling the current environment, the analyst should attempt to collect an example of each type of document. Document flow is closely tied to physical data flow. The sources or suppliers of information to the system, as well as the end users of information produced by the system, must be ascertained, and their interfaces with the system understood. It is very likely that, during analysis, the analyst will be communicating with several departments, even though a system may be generally perceived as being internal to one department. If customers, suppliers or others external to the company, interface with the current system, the analyst may gain valuable insight by talking to them. Some information regarding the shortcomings of the current system, or the importance of the information it generates, may not be obtainable from within the company, yet it may have a bearing on the proposed systems model. Following is a real example from the authors' experiences. It shows how the analyst and the company could have benefitted from discussions with customers; instead, a nightmare was created for them.

A large rubber company manufactures tires, which it sells with other automobile and hardware products to its North American customers. These customers operate service centers to which people bring their cars for servicing. As well, some customers have retail stores attached to the service centers. In these stores, people buy the products distributed by the rubber company (hardware supplies, sporting goods, etc.). This company decided that its order processing/accounts receivable/inventory control system was outdated and cumbersome, and they developed a new system to replace it.

Under the old system, a customer could order products by telephone by giving a description of the item, for example, a radial whitewall tire of a particular size and quality. The clerks on the order desk could reference the corresponding part numbers when provided with the description. Part numbers were up to twelve characters long and appeared to be random combinations of letters, dashes and numbers. Customers were sent invoices with both the descriptions and part numbers indicated, and all of the items in a shipment were printed on one invoice.

Under the new system, customers can no longer order items according to their description, but instead have to supply the correct part number. To assist them customers have been given a list with cross references to part numbers and descriptions. The list is arranged in part-number order and is more than eighty pages long.

One customer interviewed by the authors has a service center without a retail store. He has been dealing with the rubber company for more than thirty years, and had the following story to tell. He orders from 10 to 130 different items from the company per week, including tires, batteries, parts for cars, etc., and has always phoned in his orders at the end of each day. Because of his long association with the rubber company, he is familiar with the descriptions of products (but not with their part numbers). A typical part number might be X574B-WW2552. Using the new system, for each item he wishes to order, he must look through the eighty plus pages of the list supplied by the rubber company to find the correct description and its corresponding part number.

One day he telephoned in an order for twelve whitewall tires of a certain type. The order-entry clerk to whom he was speaking made an error in entering the part number and two weeks later twelve women's bicycles arrived at the customer's business. (This actually happened.) He telephoned the order department to inform them of the error and to arrange for the bicycles to be returned. He was told "of course you ordered those bicycles, the system recorded the order," and further, that the company policy did not allow returns unless the merchandise was defective. He talked to the customer accounts manager and was told the same information. He refused to pay for the bicycles, yet he could not return them. His account became overdue as a result, and the company refused to ship him more merchandise. Finally, the customer discussed his problem with a vice-president of the company, and the problem was resolved.

Another aspect of the new system is that each invoice generated contains information for one item only (one line of detail) and each is mailed individually. The customer now receives about 70 to 95 invoices per week, whereas before he received about 10. Think of the additional cost in paper, postage and handling caused by the new system, as well as the additional paperwork generated for the company's customers throughout North America!

It seems clear that the largest and most important group of users – the customers – were ignored when this system was developed. It must not have been clear to the analysts involved in the project who their end users actually were.

6.2.2 Phases 1 and 2 - Modeling the Current System

The first two phases of analysis concern modeling the current system. The product of the first phase is a set of drawings called physical data flow diagrams which describe the physical or procedural flow of data – how things are done. Physical data flow is closely tied to document flow. The first phase of analysis requires the least time of all phases, and for small projects it may be omitted entirely. If, however, the analyst is a novice, or if the analyst is unfamiliar with certain areas of the company, this short introductory step is worthwhile. When users have accepted the analyst's diagrams as accurately describing how things are done, then the first phase of analysis has been completed, and the analyst is ready to begin abstracting the logic of the physical view reflected in the physical data flow diagrams (DFD's).

In Phase 2 the analyst examines the functions of the current system, be they manual, automated or both, and using the set of physical DFD's from phase 1, abstracts their underlying logic. Phase 2 will probably involve discussions with most users of the current system and an iterative development of the logical model. Interfaces with other systems must be thoroughly understood, as well as the files used in the current system. This phase requires more time than phase 1, but relatively little time when compared to the next task, conceptualizing a new system. The main benefit of phase 2 to the analyst is that it provides the business knowledge needed for the third phase of analysis. When the user accepts the analysts' model as being accurate, the analyst can begin the most creative of all the activities, conceiving and modeling a new system.

6.2.3 Phase 3 - The Heart of Analysis: Modeling a New System

During the first two phases of analysis, the analyst models the current system, and builds the knowledge required for the real purpose of analysis: the inception and modeling of a system to satisfy a business need. This third phase of analysis is divided into two parts:

1. information gathering

2. melding the requirements obtained from the first part (information gathering) with the characteristics of the current system, to form a logical model of a solution system.

6.2.3.1 Information Gathering

In order to model an apppropriate solution to the business need, the analyst must first communicate with the potential users to ascertain in detail what is required. There are traditionally four possible methods used by analysts to gather the information necessary for modeling the proposed system.

1. the interview

2. the questionnaire

3. document review

4. observation.

We will discuss each of these methods.

The Interview
The Random House Dictionary defines an interview as "a meeting for obtaining information by questioning a person." The success or failure of an interview in the context of analysis depends largely on the analyst's preparation and abilities with respect to listening, verbal communication and objectivity.

During the preliminary stages of analysis, the analyst should concentrate on obtaining a general, high-level view of what is required by the new system. This will entail gathering qualitative information such as the importance, purpose and objectives of the new system from the corporate view, and the priorities within it. Management are the people who can provide such information; from them the analyst will receive information that is otherwise unobtainable or undocumented.

Early in Phase 3 the analyst should prepare a set of questions to help in obtaining the overall view. The set of questions should start with the general, leaving more specific questions to later. The function of the questions is to ensure that all aspects of the topic or area about which the analyst desires information are covered. Interviews which take place during the initial stages are for providing qualitative information. As the analysis proceeds, however, the investigation becomes increasingly detailed, and the interviews will have more structure, centering on a detailed set of questions.

A point worth remembering when preparing a set of questions is that questions should be eliminated whenever the information is obtainable by another method, for example, by document review. To ask unnecessary questions wastes the time of the respondent, and may also influence the success of the remaining portion of the interview if the respondent perceives the analyst as being unprepared and using time that would be better spent elsewhere. The interview will be successful only if the respondent wants to communicate, and one of the analyst's goals should be to engender this willingness in the respondent [12].

The Questionnaire

In analysis, interviews are an essential method for obtaining information, especially during the early stages when the information has a qualitative nature. However, as the analysis proceeds, the nature of the information required becomes more quantitative, and sometimes a questionnaire can be useful for this purpose. As well, questionnaires may be effective when information from a large number of people must be obtained. Standardized questions requiring quantitative answers can provide the analyst with reliable information.

The questionnaire should be designed with a particular objective in mind. The questionnaire in Figure 56 was created to obtain information about the usefulness of, and preferred formats for, customer statements. The questionnaire was mailed to each customer of an electronics company, First City Computers, along with the customer's statement (Figure 57) and a self-addressed, stamped envelope.

Document Review

Reviewing company documents such as inventory reports or tickets, invoices, customer statements, purchase orders and financial reports may provide the analyst with quantitative information that would otherwise have to be obtained by interviewing or by questionnaire. Standard procedures manuals or specifications of systems used throughout the company can usually furnish additional qualitative information. Document review is a secondary method of information gathering which complements the interview and the questionnaire.

Observation

There are often situations in which observation of people doing their jobs is essential to the analysis of a system. For example, if you are an analyst employed in a stock brokerage firm and the new system concerns traders who work on the trading floor, then it will probably be necessary to observe them doing their jobs. Or, suppose you work at a manufacturing company where the

We are developing a new Customer Accounts computer system at First City Computers, and wish to consider our customers' opinions in the process. We would like some information from you regarding the customer billing statements we mail each month.

Please complete this questionnaire and return it to us in the enclosed envelope within 30 days.

1. Do you use the statement?

2. If your answer is no, please explain.

3. Do you find the top portion of the statement that has the name and address information easy to read and understand?

4. How could we improve the top portion of the statement?

5. The middle or detailed section of the statement gives information about the balance forward from the previous month, as well as listing each invoice, credit note and resulting balance as events occur throughout the month. Is this information easy to read and understand?

6. If your answer to question 5 is no, please explain.

7. How could we improve the middle section of the statement?

8. Information summarized at the bottom of the statement includes the amount you paid this month and the ageing of your account. Is this information useful?

9. What other summary information not shown now would be useful to have?

10. Do you have any other suggestions concerning the statements?

11. Indicate below which items apply to you.

    Private Customer
    Distributor
    We buy less than $5,000 each year
    We buy between $5,000 and $10,000 each year
    We buy more than $10,000 annually

Figure 56: First City questionnaire

FIRST CITY COMPUTERS

ACCOUNT NUMBER: QUA5000 DATE: June 30

Quality Business Services
1223 North Ridge Road
Lancaster, Pennsylvania
32110

DATE	ITEM	INVOICE	AMOUNT	BALANCE
June 1	Balance Forward		125.40	
June 5	payment	467	25.40	100.00
June 11	invoice	901	759.12	859.12
June 16	invoice	2260	1,742.00	2,601.12
June 22	credit note	300	50.00	2,551.12
June 25	payment	901	759.12	1,792.00

Payments and Credits: 834.52

Current: 1,692.00 Over 30: 17.85 Over 60: 82.15 Over 90: 0.00

Balance Due: 1,792.00

This is a statement of your account. Accounts are payable on the tenth of the month following.

Figure 57: Example of First City customer billing statement

plant employees work at work stations and operate various pieces of equipment such as punch presses and lathes. If you are developing a cost accounting system for the company's products, observing the plant workers at their job will help you to understand the labor costs involved in the manufacturing process. (The analysts involved in the development of the order processing system of the rubber company could have benefitted by manning the telephones at the order desks, for they would have learned the kind of information customers actually supply when placing orders with the company.)

6.2.3.2 Incorporating the Information into the New Systems Model

The four methods discussed above are used for obtaining information. After the analyst has gathered the necessary information, the task of conceiving and logically modeling the new system begins. The information should be assembled into a profile of the new system requirements starting with the most general and concluding with the most specific. The basis for this profile, of course, is the set of objectives coming from the requirements analysis document. The analyst will have discovered more detailed objectives during the recent information-gathering activities.

Sometimes there will be contradictions among the objectives. When this occurs, the analyst should choose those that are compatible with the corporate objectives, or, discuss the anomalies with the manager responsible for the project.

6.2.4 Alternative Approaches to the First Two Phases

Some corporations use an alternative method for deriving the logical design of software. The most popular of these alternatives was developed by IBM in 1977, and is called joint application design, or JAD. With JAD, or JAD-like methods, the requirements determination phase of the traditional life cycle and steps 1, 2 and 3 of analysis are combined to form a single phase. The logical design of the software takes place during one or several, intensive two- to four-day workshops attended by both users and systems people.

The objectives of this method are the acceleration of the development process and the improvement of the quality of logical design through increased user involvement. Through successful applications of this type of method, many companies have been able to reduce the cost and time of development and to increase the quality of the resulting system, relative to traditional methods.

The basic approach is to conduct intensive, structured workshops directed by a person experienced in leading such sessions. The leader uses materials tailored to the particular application to guide group members through the steps necessary to logically design a system. The JAD method was developed for use in small, on-line transaction-based operational systems; however, it may be altered to suit information-centered systems by emphasizing the information needs of the users.

The emphasis in the workshops is on the flow of work through the area under study. Materials are designed to focus the group's attention on business activities, data elements and screen formats. Participants typically include a trained session leader, several potential users, one or two systems people (usually the project leader and possibly one other), perhaps one or two representatives from other areas of the company and one or two "scribes." The participants are chosen by management and collectively represent expertise in all aspects of the system under investigation.

During each session, users describe their business functions and information needs, the data elements they use, and their interface with the system. They also describe how they want the new system to affect their business functions.

The function of the scribes is to document the proceedings and the decisions of the group. Normally one scribe will be from the systems group and the other from the user community. The scribes recap each decision and at the conclusion of a session they consolidate the group's decisions into a report. The group has the authority for all decisions regarding the system except those requiring knowledge or authority beyond their scope.

Under the direction of the leader, all aspects of the application are addressed. The result, after a number of sessions, is the logical design of the software solution.

Companies have experienced both success and failure with JAD-like methods. Some advantages of JAD appear to be

1. A more user-oriented system than otherwise would have been obtained, thereby reducing future maintenance costs.

2. Reduction in development costs and time. (Studies have shown development time reductions of up to 40 percent.)

3. Increased communications and understanding between users and data processing personnel.

Some potential disadvantages may include

1. Difficulty in obtaining and training session leaders.

2. Difficulty in the application of the method to large complex projects.

6.2.4.1 Alternative Approaches: Conclusions

During the first two phases of analysis, the analyst models the current system. The third phase of analysis, deriving a model of the new system, is divided into two parts:

1. information gathering

2. melding the requirements obtained from the first part (information gathering) with the current way of operating, to form a logical model of a solution system.

Traditional techniques used in the information-gathering process are the interview, the questionnaire, document review and observation.

JAD and other related methods have been successfully used as alternatives to the first portion of the traditional life cycle. After the logical design has been produced, the remaining phases of the TLC, as discussed in Part 1 of this book, can proceed. For some applications and in some business environments, these alternative methods can be a cost-effective way for both shortening the development period and producing a better product.

6.2.5 The Last Three Phases of Analysis

The last three phases of analysis (4 through 6) concern the development of possible automation strategies, management's selection of a strategy, and preparation of the selected strategy for the design phase. First, several strategies for automation, along with a cost/benefit analysis for each, are developed and presented to management. Next, management select one of the options (with or without revisions). Finally, the agreed-upon solution is packaged for final approval and presentation to user and systems management. Part of this package, called the function specification, provides the documentation input to the design phase of software development.

6.2.6 Phase 4 - Developing Strategies for Automation

After the first three phases have been completed, the analyst must develop several possible strategies for automation. One strategy may be a totally ideal system comprising full automation of all processes described in the system analysis. The other extreme is a system that remains as it currently is. This "status quo" solution means no use is made of the analysis. Depending on the complexity and magnitude of the business function analyzed, there are normally many possible automation strategies between these two extremes.

Normally the full-blown idealized system is much too large and complicated to be implemented without modification. It may require years to complete and its cost may be prohibitive. There are several possible ways in which it can be partitioned into automation options. One approach would be to partition along business function lines, with those features of the idealized system that pertain to certain departments or business functions in the company forming separate modules. This approach creates several modules whose union is the idealized system. A specific module can then be chosen for design and implementation, while the others may be scrapped or postponed.

Another more common approach, is to partition the idealized system into modules defined according to urgency and potential benefits. For example, one often finds that certain features of a system are essential (otherwise the system is not useful), while others are desirable or "nice to have." The latter can be dispensed with now and added later. Some functions may clearly be needed immediately, while others are needed but can wait.

Following are two examples of such decisions:

1. In Chapter 5 we described a decision support system based upon an idealized concept comprising an information system, a simulation program and a forecasting function. This project was ultimately implemented by partitioning into two modules. The first module consisted of the information system requirement (dealing with historical and competitive data) and the simulation program. The second module allowed for forecasting. The decision to partition in this fashion was based on urgency. The information system and simulation program were deemed much more important than the forecasting function. Another reason for this decision was that relying on forecasting was deemed risky, hence if anything was to be postponed, forecasting was the obvious choice.

2. A major bank had a trading room consisting of several dozen people who traded, bought and sold stocks, bonds and money-market instruments. The trading room was required to automate trade processing and provide decision support tools. A systems analysis resulted in a total, idealized concept deemed too large to implement at one time. The solution was to automate each product type's trading desk in separate modules, one after the other.

6.2.7 Phase 5 - Selection of a System

Usually idealized concepts are too large to implement fully and often they must be partitioned into manageable modules, or sometimes reduced to one final subset. In order to arrive at a decision about which system to implement, the different options must be described in relation to the whole (ideal system). Data flow diagrams and corresponding brief written descriptions are adequate in this regard.

It is important to provide a rough cost/benefit analysis and an indication of the business impact of each option. Also, a rough implementation schedule for each should be provided, showing milestones and finished product dates.

The final recommended approach, be it a series of modules or just a one-time development, is made by both the user and systems communities in consultation with, and presentation to, responsible management. Changes are often made during this process and sometimes diplomacy is required because disagreements over the necessity of specific features can occur.

Once a final solution is agreed upon and approved by the appropriate levels of management, one last decision remains: whether to develop the system in-house or to purchase an existing software package (which itself may require enhancing). In many cases a package already exists that performs at least 75 percent of the required functions. It may be available from a software vendor or from another company who previously developed the system in-house and who wish to recover some of the development cost by selling it.

To aid in the make-buy decision, the team normally contacts other companies, ascertaining whether or not they use a packaged system, and if so, what their assessment is. Often someone knows of existing packages. More often a formal approach is used whereby each appropriate vendor is sent a request for information (RFI).

An RFI, as the name implies, is meant to determine whether or not a vendor sells packaged software capable of providing the final solution. It consists of a covering letter explaining the purpose of the RFI, outlining the company's requirements and setting a deadline for replying. RFI responses usually include descriptions of any relevant packages, their cost schedules, estimated delivery dates, hardware requirements, company profile and client references. Normally costs are not binding – they are merely estimates for comparison. An RFI can actually be sent early in analysis, before the final solution has been specified, to save time.

Replies to the RFI's are evaluated by the team, and a short list of vendors chosen. Alternatively, a decision to develop the system in-house may be made. If the decision is to pursue packaged software, the short-listed vendors are each sent a request for proposal (RFP). (Note that some companies call this a request for quotation, or RFQ.)

The RFP is much more detailed than an RFI. Its core is the detailed functional specifications for the final solution, including DFD's, data dictionary updates and specifications for each process within the new system (process logic). It also contains the following:

1. An objective and scope

2. A description of the company and the reasons for automation

3. A section on expected responses to the RFI covering

   - deadline
   - sample contract
   - costs
   - implementation schedule and plan
   - hardware and operating system requirements
   - personnel and other operating costs
   - maintenance contract and costs
   - list of customers using the product
   - description of vendor company and its financial health
   - detailed description of proposed solution

4. Functional requirements

5. Response time, security and error-recovery requirements

6. Contact persons for clarification of questions

7. Evaluation criteria.

The costs quoted are normally binding for 30 to 90 days, and the selected vendor will sign a contract with the company for the expected product and timetable.

For complex or unique applications, a package may apply only after substantial enhancements. The cost of modifying the package will be detailed in the response to the RFP.

Again the team evaluates responses. The criteria for evaluation have been given to each vendor and usually include:

1. Costs
2. Delivery dates
3. Closeness of fit to requirements
4. Evidence that the vendor can meet any deadline specified and deliver a quality product.

Once all responses are evaluated and a suitable vendor selected, the team may work to compare the make or buy scenarios. At this point a final solution is selected – either to develop in-house or to purchase. Once that decision has been made, a more detailed cost-benefit analysis and implementation plan is prepared and the final solution is packaged.

6.2.8 Phase 6 - Packaging the Selected System

The final phase in analysis is the documentation of the chosen solution. The title given to the resulting document contains the name of the project and the subtitle given is "Functional Specification and Business Case." This document is described in detail in Chapter 9.

Finally, management (usually, but not always, the user management) must use this document to make a business case (i.e., costs and benefits) in order to obtain corporate approval for assembling the human and financial resources to implement the system.

6.3 Knowledge and Skill Requirements

Systems analysts have one of the most interesting and important positions in the systems division of a company. On the one hand, they work with computers and systems, something they enjoy and for which they have an aptitude. On the other, they are involved with people, from both user and systems areas, and work on specific business problems facing the company. Their ability to create effective systems is of ever-increasing importance as more strategic use is made of information systems. The variety of applications and the need for creative solutions to business problems makes this a challenging and exciting position.

Because systems analysts must bridge the gap between business functions and computer systems, they require a fairly broad range of knowledge and skills. In this section we discuss the areas of knowledge and the particular skills that a systems analyst should either possess or develop to become as successful as possible in this career. The topics covered are divided into technical and business areas.

6.3.1 Technical Knowledge and Skills

Technical *knowledge* necessary for system analysis is basically that knowledge concerned with computing:

- Programming
- Data Structures
- Database Design
- Systems Design
- Tools of Analysis (Chapters 7 & 8)
- Computer Hardware
- Data Communication.

Other than the tools of analysis (data flow diagrams, decision tables, decision trees and structured English), these knowledge areas may or may not be utilized directly during analysis. However, they are very important generally, in terms of knowing how systems are designed and built, what is feasible and how much effort is required to build a system. Knowledge in these areas gives the systems analyst insights, insights that may make the difference between mediocre and high quality results. Knowledge in all of these areas can be gained through courses or experience.

Technical *skills* required for systems analysis are:

- Analytical
- Conceptual
- Quantitative.

Analytical skills are an obvious requirement for system analysis, in which a process, function, aspect or subject is separated into simpler, more easily understood parts. Analytical skills are applied when a systems analyst undertakes to understand the current business situation. This is, in fact, the first stage of modeling the processes to be automated.

The ability to conceptualize is needed in all phases of analysis, especially during the phase of creating a systems solution to the perceived problems and then creating a logical model of the required system. Overall, the skill of conceptualization is needed in order to "see the forest, not just the trees."

Development of decision support systems and systems involving analytical modeling (systems that analyze business trends and predict the results of changes – "what if" analysis) is increasing. These systems have their basis in mathematical methods, and there is an obvious need for the systems analyst to possess quantitative (basic mathematical and statistical) skills for understanding such requirements. The same skills are needed in cost/benefit analysis, and for estimating project costs and resource requirements. Quantitative skills are closely tied to analytical skills in that they both require a firm grasp of the economics and mathematics of a problem.

The technical knowledge and skills outlined above are fairly obvious requirements for effective systems analysis. Perhaps less obvious are the required business skills. These are discussed in the next section.

6.3.2 Business Skills

Much of the technical knowledge, and many of the skills pertinent to systems analysis, can be acquired through formal education. On the other hand, business skills are not usually included in a person's academic education; they must be mastered through direct business experience and conscious effort, or through special outside business training. These skills are of great importance in today's corporate environment, and of especial importance to systems analysts.

The need for business skills arises because systems analysis deals with business problems, more importantly, with the people who identify, understand and need solutions to these problems. The systems analyst must gain an understanding of the business area, its processes, problems and relationships to other areas, and must be able to communicate this understanding to other systems people and to management. When solutions are developed and systems proposed, the systems analyst must be able to describe complicated ideas to non-technical people (users) and resolve disputes or misunderstandings. The following business communications skills are required by the analyst:

- Oral communication
- Listening
- Interviewing
- Written communication
- Presentations
- Diplomacy
- Teamwork.

Oral communication, listening and interviewing skills are probably the most important skills for a systems analyst to possess. These are required during all phases of analysis. Some typical situations where these skills are needed are as follows:

- When an analysis project begins, the analyst may have to orient users who have had no previous experience in analysis. This task will require an explanation of what analysis is, why the various steps are necessary, how the process is to be carried out and what the end result will be. It is important to give concise explanations, both of the process itself and of the benefits to be derived, without making motherhood statements about solving all of the users' problems. The goal is to make users understand what is required of them and the other team members.

- During analysis, the analyst will be striving to gain an understanding of the business area in question. Most of this understanding will be obtained through discussions with managers and staff and will require the ability to listen to others, and to explain what is to be learned and why it is necessary to learn it. This is the time during which interviewing skills are especially valuable.

- Meetings are another fairly regular occurrence in the analysis process. These meetings can consist of the team alone, or of the team and management, with or without user participation. It is very important to listen carefully and to speak effectively in such meetings, otherwise valuable time can be wasted and credibility lost. As a general rule, most people dislike meetings, even though they can be important. This may be due to communication problems, lack of agenda or not knowing the purpose of the meeting.

In general, these interpersonal communication skills are not easily taught. Rather, they are developed through effort, commitment and practice. In the case of interviewing, however, we can provide some guidelines:

- Schedule each interview ahead of time, as you would any appointment, and inform the respondent of your purpose. This will allow the respondent to prepare for the interview.

- Learn something about the respondent in advance. This information may be used for "breaking the ice," and may put both you and the respondent at ease.

- Commence the interview by introducing yourself. Give a brief outline of your purpose and the importance of the project.

- During the interview listen carefully and learn by what the respondent is saying. Allow respondents to express their own views. You should follow up on points or issues raised in order to gain insight into new considerations. You should be guided by your own common sense; for example, if as the interview progresses, you feel that the original questions are no longer appropriate, then ask new questions as opportunities arise.

- Take a minimum of notes. The time is better spent listening.
- Just prior to terminating the interview, give the respondent a chance to mention any important points that might have been omitted.
- At the end of the interview, you should verbally summarize the information obtained. As well, a written summary should be prepared and, if appropriate, sent to the respondent with a request for comments from the respondent about the summary.

Formal presentations are another regular occurrence during analysis, at least for large projects. These are given at the initiation and completion of a project, and at milestones to apprise management of progress. At formal presentations, verbal skills are truly tested. However, even a good communicator can deliver an ineffectual speech because of nervousness, and unfortunately, a poor delivery can jeopardize an otherwise excellent proposal or idea. On the other hand, a polished delivery can make a superficial idea look marvellous. Systems analysts who make every effort to develop the skills required for giving effective presentations are a credit to their project, to their company and to themselves.

There are courses and organizations (e.g., Toastmasters International) devoted solely to the development and maintenance of presentation skills. As with most skills, practice and commitment are required in order to achieve reasonable proficiency. Some pointers for making effective presentations follow:

- The presentation must be at the correct technical level and contain the appropriate degree of detail for the target audience.
- The purpose of the presentation must be clearly stated (is it for information only or is a decision to be made, etc.).
- An outline of the talk should be presented.
- The presentation should be logically organized, with a beginning, a middle and an end. The end should contain a conclusion.

- State whether or not questions may be asked during the talk or saved until it is finished.

- Audio-visual aids are normally used (handouts are helpful but can be distracting – give them out afterward).

- Overhead transparencies or slides must be simple (key words or points).

- Face the audience, not the screen or board.

- Avoid conversational static (uhm, uhh, like).

- Avoid jargon and unnecessary details (you are there to inform, not to impress).

- Rehearse mentally or physically, but do not memorize.

- Be prepared to cut points if you exceed the allotted time.

These are just some of the more obvious points to keep in mind when preparing and delivering a presentation. They may be more relevant after you have experienced some good (and not so good) talks.

Writing skills are one aspect of verbal communication skills. Writing is the one skill related to business that is taught in all levels of formal education, and it is one that improves through practice. During analysis this skill is brought to bear in the preparation of reports (status and progress reports), RFP's and especially in the preparation of the functional specification and the business case. The business case is described in detail in Chapter 9.

Points to remember for writing are similar to those for making presentations:

- Reports or documents must be logically organized.

- Objectives or purposes should be clearly stated.

- Place necessary support material in an appendix if possible.

- Avoid unnecessary details.

- Writing style should be clear and free from jargon.

- Bear in mind who the target audience is (management, users, peers).

Diplomacy is a skill often required when there is disagreement among team members or between user and systems management. A typical situation consists of users asking for a requirement and systems people thinking it unnecessary, infeasible or too expensive relative to its benefits. Depending on the problem, the analyst may have to convince one party that the other is correct or that the other party's desire is to be respected. This is not always easy, and often management is brought in to settle the dispute. If, for example, you as the analyst have determined that a particular requirement is not feasible you must be very careful not to just tell the users that they are wrong, but rather lead them through the steps you took to determine the infeasibilty. It is rare for a large project to generate no disputes.

Since most systems analysis is accomplished in projects, it is clear that the ability to lead and to work on a team is necessary. An effective leader must ensure that all team members know exactly what is expected of them in terms of work and their roles relative to the whole project.

6.4 Exercises

1. Discuss the importance of the analyst's ability to listen, to communicate orally and to be objective in an interview.

2. Suppose you are an analyst working for the rubber company described in this chapter. The company has decided, after many complaints from its customers, to develop a new order processing/accounts receivable/inventory control system. You are involved in this project, and one of the objectives of the new system is to satisfy customers – to allow them to place orders over the telephone by product description, to produce multi-item invoices, etc. You believe that sending a questionnaire to customers is appropriate. Because of the large number of customers, you will select a sample size of 10 percent to receive questionnaires. How do you decide who should receive questionnaires? Construct an appropriate questionnaire to send to customers.

3. Suppose that as an analyst you are interviewing a person who firmly dislikes the current system. Should you play the role of counselor, asking what is wrong with the current system and what can be done to correct it? Why?

4. Suppose you are interviewing a person who is afraid of being made unemployed by a new system. You have to interview the person because the information you require is not obtainable elsewhere. What can you do to diffuse potential problems and obtain the most information?

5. Suppose you must interview a person and have prepared a set of questions. Although some of the questions may threaten the respondent, they must be asked. When during the interview should they be asked? Why?

6. Suppose that during an interview, you receive an answer that you do not understand. What should you do?

7. Develop several ways in which, during an interview, you can help respondents expand on what they have just said.

8. Suppose you have to observe an employee working in order to obtain information essential to the new system. The person feels uncomfortable with you watching. What could you do to help put the employee at ease?

9. Discuss the potential difficulties in applying a JAD-like approach to a large, complex project.

10. What role do you see the requirement specification document playing in the last three phases of analysis?

11. How can a company's business objectives help in determining strategies for automation?

12. The bank trading floor example indicated automation along product lines (e.g. bonds, then money market, then stocks). What other strategies are possible?

13. Suppose you were responsible for automating a chain of retail stores. Functions typically automated are:

    - Inventory
    - Cost/revenue
    - Cash register functions (item entry, sales tax computation, change, etc.)

- Payroll

All of these can be treated independently but there is a clear advantage to an integrated approach, for example, payroll with cost/revenue, cash register functions with inventory, inventory with cost/revenue, etc. There are several automation options possible:

- Automate one store completely, then do the others.
- Automate all stores together along functional lines, for example, automate completely all inventory, then cost/revenue, etc.
- Automate several functions together in all stores, then other functions.

What are the benefits and disadvantages of each of these different approaches? Can you create others?

14. What are the pros and cons of packaged software solutions? Include risks.
15. What skills do you see as being most important for systems people assigned to help in a requirements analysis? Support your answers.
16. Why should solutions not be stressed during the requirements phase?
17. In computerization or automation, what are the underlying reasons for doing a requirements analysis?

6.5 Bibliography

1. Anthony, Robert N. and Welsch, Glenn A. *Fundamentals of Management Accounting*. Homewood, Illinois. Richard D. Irwin, 1977.
2. Barker, Larry L. *Listening Behavior*. Englewood Cliffs, New Jersey. Prentice-Hall, 1971.
3. Biggs, Charles L., Birks, Evan G. and Atkins, William. *Managing the Systems Development Process*. Englewood Cliffs, New Jersey. Touche Ross and Company, Prentice-Hall, 1980.

4. Brooks, Fred P., Jr. *The Mythical Man-Month: Essays on Software Engineering*. Reading, Massachusetts. Addison-Wesley, 1975.

5. Brown, David B. and Herbanek, Jeffrey A. *Systems Analysis for Applications Software Design*. Oakland, California. Holden-Day, Inc., 1984

6. Cash, James I., Jr., McFarlan, F. Warren and McKenney, James L. *Corporate Information Systems Management: Text and Cases*. Homewood, Illinois. Richard D. Irwin, 1983.

7. Curtis, Bill (editor). *Tutorial: Human Factors in Software Development*. Los Angeles, California. IEEE Computer Society, 1981.

8. DeMarco, Tom. *Structured Analysis and System Specification*. Englewood Cliffs, New Jersey. Prentice-Hall, 1979.

9. DeMasi, Donald. *An Introduction to Business Systems Analysis*. Reading, Massachusetts. Addison-Wesley, 1969.

10. Donaghy, William C. *The Interview: Skills and Applications*. Glenview, Illinios. Scott, Foresman and Company, 1984.

11. Gane, Chris and Sarson, Trish. *Structured Systems Analysis: Tools and Techniques*. Englewood Cliffs, New Jersey. Prentice-Hall, 1979.

12. Gildersleeve, Thomas R. *Successful Data Processing System Analysis, Second Edition*. Englewood Cliffs, New Jersey. Prentice-Hall, 1985.

13. *Guide 64*. Session no. MP-7460. Los Angeles, California. IBM Corporation, March, 1986.

14. *Guide 60*. Session no. MP-7436B. San Francisco, California. IBM Corporation, November, 1984.

15. Hodge, Bartow and Clements, James. *Business Systems Analysis*. Englewood Cliffs, New Jersey. Reston Publishing, 1986.

16. Langs, Robert. *The Listening Process*. New York, New York. Jason Aronson, 1978.

17. Martin, James. *Managing the Data Base Environment*. Englewood Cliffs, New Jersey. Prentice-Hall, 1983.

18. McFarlan, F. Warren and McKenny, James L. *Corporate Information Systems Management: The Issues Facing Senior Executives*. Homewood, Illinois. Richard D. Irwin, 1983.

19. Parkin, A. *Systems Analysis*. London, U.K. Edward Arnold, 1980.

20. Parkin, A. *Systems Management*. London, U.K. Edward Arnold, 1980.

21. Pressman, Roger S. *Software Engineering: A Practitioner's Approach*. New York, New York. McGraw-Hill, 1982.

22. Rockart, John F. and Bullen, Christine V. (editors). *The Rise of Managerial Computing: The Best of the Center for Information Systems Research*. Homewood, Illinois. Dow Jones-Irwin, 1986.

23. Selig, Gad J. *Strategic Planning for Information Resource Management: A Multinational Perspective*. Ann Arbor, Michigan. UMI Research Press, 1983.

24. Senn, James A. *Analysis and Design of Information Systems*. New York, New York. McGraw-Hill, 1984.

25. Sprague, R.H. and Carlson, E.D. *Building Effective Decision Support Systems*. Englewood Cliffs, New Jersey. Prentice-Hall, 1982.

26. Sprague, Ralph and McNurlin, Barbara. *Information Systems Management in Practice*. Englewood Cliffs, New Jersey. Prentice-Hall, 1986.

27. Steil, Lyman K., Summerfield, Joanne and de Mare, George. *Listening: It Can Change Your Life*. New York, New York. John Wiley & Sons, 1983.

28. Thomsett, R. *People and Project Management*. New York, New York. Yourdon, Inc., 1980.

29. Weinberg, Gerald M. *Rethinking Systems Analysis and Design*. Boston, Massachusetts. Little, Brown and Company, 1982.

30. Weinberg, G.M. *The Psychology of Computer Programming.* New York, New York. Van Nostrand Reinhold, 1971.

Chapter 7

SYSTEMS MODELING USING DATA FLOW DIAGRAMS

A model is a representation of something. Its purpose is often to show the components or structure of an object. In the fields of architecture and engineering, models may be built from metal or plastic, developed using a computer (computer-aided design) or produced with the aid of other suitable tools and materials.

Systems analysis concerns the process of conceptualizing a software solution to a business need, then modeling the solution and using the model in subsequent phases of development. Software systems may be modeled from different perspectives. One way is modeling according to the structure of data [10] so that the structure of data is reflected in the ultimate structure of the software using Jackson diagrams or Warnier diagrams [2]. The basic premise of these methods is that the structure of the data is more stable than the procedures that use the data.

Another standard method of modeling software systems is according to the functions performed within. Data flow diagrams are a tool used for this purpose; they depict a business area according to the functions performed on the data.

In the next two chapters, we present an explanation of data flow diagrams and other complementary tools, together with their reasons for being and an indication of where they are used during analysis activities.

7.1 Purpose/Use and Definition

A *data flow diagram* (DFD) is a network drawing that describes the flow of data through a system. Included in a data flow diagram are all external systems that originate or receive data, the transformations or functions that process the data, and the places in the system where data are stored.

Data flow diagrams are used in several phases of analysis. They model the current business area under investigation. DFD's also model the software solution of the user's business needs, the system conceived by the analyst that will be automated. These particular DFD's are used in subsequent phases of software development. During the design phase the conceptual model produced during analysis is transformed to a physical model, by standard methods used to effect the transformation [18]. By their nature, data flow diagrams allow decomposition of a system according to function, and these functions form the basis of the program modules coded during the programming phase of the life cycle.

DFD's are easy to read, and because they are graphic, they are concise and precise. They help greatly to see the "forest" as well as the "trees." Data flow diagrams are a non-technical and valuable tool for user-analyst communications. The analyst and user can use the DFD model as a basis for common understanding and agreement, for both the present environment and any proposed system.

7.2 Where They Fit

Data flow diagrams are used in several phases of analysis. In the first phase, when the user's current physical environment is modeled, a *physical* data flow diagram is produced. During the second phase of analysis, once the user and analyst have agreed that the DFD's from the first phase represent *how* things are currently done, the analyst abstracts the underlying logic in the current system to develop a *logical* data flow diagram. Logical data flow diagrams show the functions in a system. Still later in analysis, the analyst conceives a logical solution of the user's needs and models this solution. The data flow diagram of the solution system is called a *new* logical data flow diagram.

After modeling a solution to satify the user's business requirement, the analyst develops several possibilities for automation and presents them to management. After management have selected one of the options, the data flow diagram of that option is packaged, along with other documentation, in preparation for subsequent phases of the traditional life cycle. This documentation, called the functional specification, is input to the design phase that follows analysis.

7.3 Components

Data flow diagrams are developed and used throughout analysis and represent and important tool to the systems analyst. In order to demonstrate the components, conventions and rules that apply to DFD's, we will start with the data flow diagram of the steps of analysis from Chapter 6 (Figure 58). Then we will develop a set of physical and logical DFD's using First City Computers as our example. Finally, having developed data flow diagrams, we will summarize our results.

1. *Sources/Sinks*:

   Any system, entity or function external to the area under consideration but interacting with it (actively or passively) is either a *source* of data or a *sink* of data and will denoted by a rectangle.

   Following the example given in Figure 58, the process of analysis really starts with the user, who is not part of the system under consideration, but exists independently. The user is an originator or *source* of data, and interfaces with our system (analysis). The requirements determination and design phases of software development are also independent systems. The requirements determination phase is a data source. The design phase is a *sink* or receiver of data. Sources and sinks interface with the system under study through data flows and data files.

2. *Data Flows*:

   The named arrows on DFD's are the *data flows*. A data flow is a grouping of data that moves together within a system in the direction indicated by the arrow. Data flows may be communicated by letters, memos, telephone, telex, etc. A data flow is an interface between other components of a DFD.

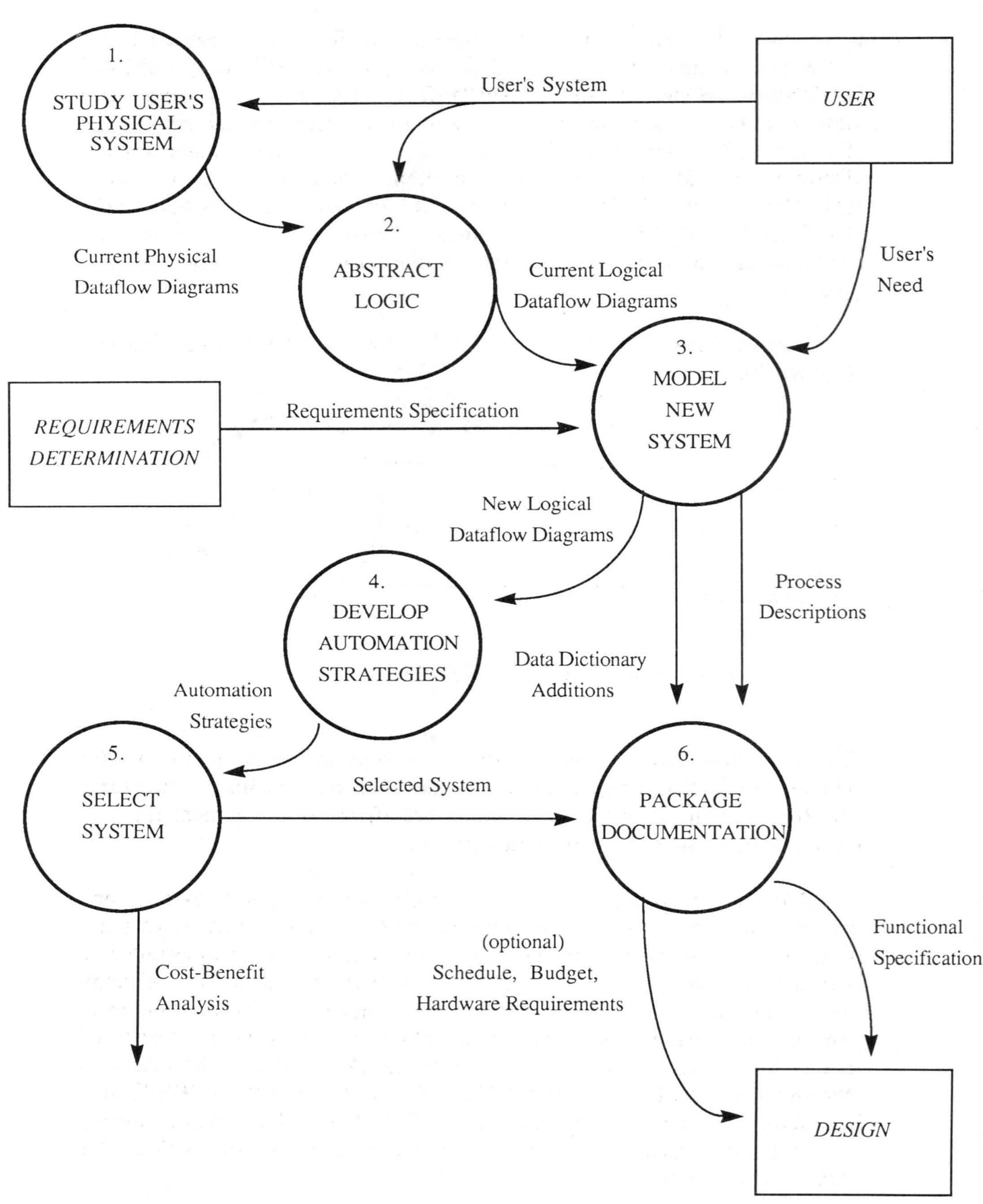

Figure 58: Steps of analysis

Figure 58 contains many data flows. The flow User's System flows from the source *USER* to the two processes STUDY USER'S PHYSICAL SYSTEM and ABSTRACT LOGIC. It is a *diverging* data flow because it has more than one destination. Except for diverging data flows, data flow names must be unique. The two data flows called Process Descriptions and Data Dictionary Additions are output from process 3 (MODEL NEW SYSTEM) and input to process 6 (PACKAGE THE DOCUMENTATION). They are shown separately, even though they have the same source and destination, because they are separate data groupings.

Consider now an excerpt from a DFD of an accounts payable function (Figure 59).

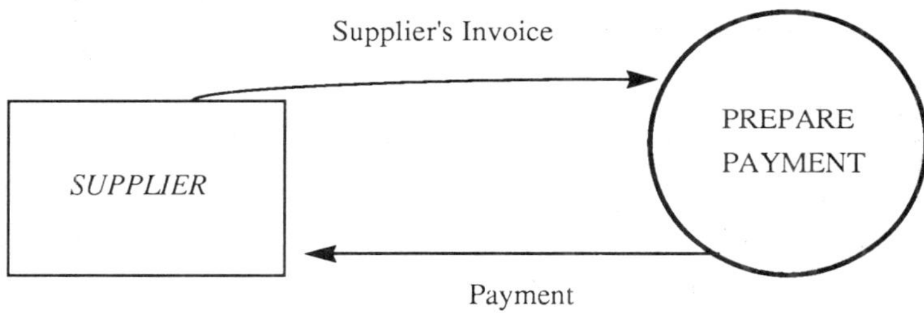

Figure 59: Excerpt of accounts payable DFD

The data flow Payment may consist of a copy of the supplier's invoice plus a check. Even though there is more than one document composing the flow, together they form the single data flow called Payment. Flow of data is not the same as flow of documents.

Flows of control, flags and *prompts* to activate processes are commonly used terms occurring in computer programming. Normally they are data items or program variables used for a special purpose: they indicate the logic to be performed. Their data content per se is not relevant. DFD's do not show flows of control, time sequences, flags or prompts. The following example illustrates what DFD's do *not* show. In Figure 60 notice the portion of the DFD in which the flow End Of The Month is shown going into the process PRODUCE CUSTOMER STATEMENTS. The diagram is incorrect because End Of The Month is a prompt, causing customers' statements to be produced, but containing no information. This data flow should not appear on the DFD.

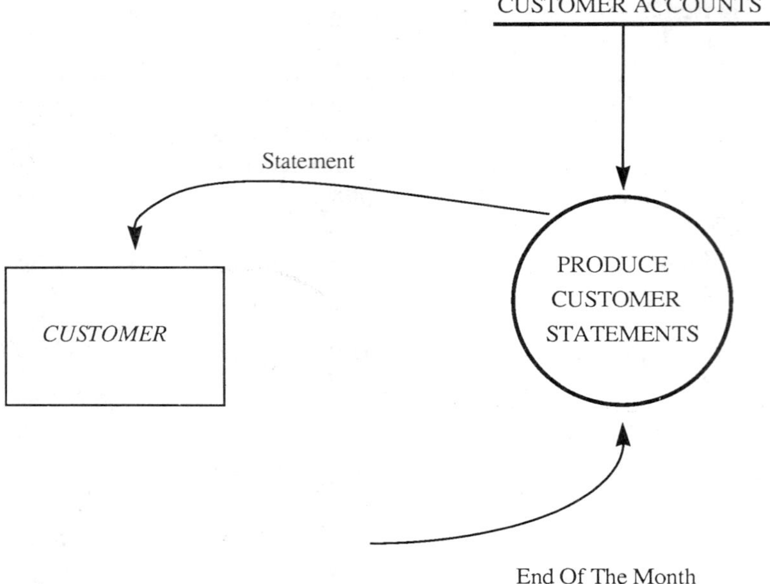

Figure 60: Incorrect DFD

3. *Processes/Functions*:

The labeled circles on a data flow diagram are the *processes* or *functions* that operate on, or transform, the data. The processes may be manual, automated or a combination of manual and automated (mixed). Notice the similarity to the mathematical concept of *function*, which also transforms, or operates on, data. Figure 61 illustrates how processes are described at the physical level.

MARY receives payments from customers and deposits the day's receipts at the bank.

The logical equivalent of the process MARY in Figure 61 would be the excerpt that appears in Figure 62. Mary is involved in several distinct logical processes: she applies payments to customers' accounts, enters them into the cash receipts journal and prepares the bank deposit.

The two processes UPDATE RECEIPTS and UPDATE CUSTOMER ACCOUNTS, use the same data flow, Payment, which diverges to both of them. In order to keep track of complicated situations requiring large numbers of processes, each process in a DFD has an iden-

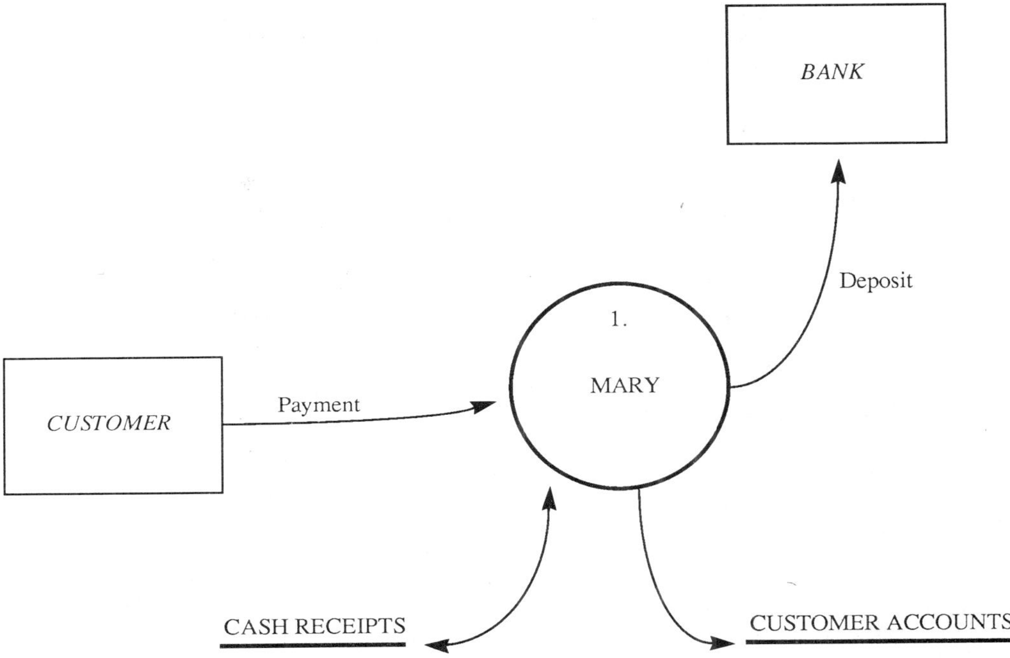

Figure 61: Excerpt of physical DFD

tifying number. Numbering conventions for DFD's will be explained later in the chapter.

Figure 58 contains six processes. Diagrams having too many processes are not easily read. A general guideline for drawing data flow diagrams is to limit the number of processes shown on one diagram to seven or eight.

Physical DFD's depict *how* functions are performed, not actually *what* is being done. Logical data flow diagrams, on the other hand, depict *what* is being done. For example Figure 61 shows MARY as a process, but provides no information about the business functions Mary performs.

4. *Data Files*

Data files are shown by named straight lines. Remember that these are files at a conceptual level and indicate any place where data resides: a computerized database, telephone directory, filing cabinet, price list, or a person's memory. Data files correspond to entities in the entity-relationship model.

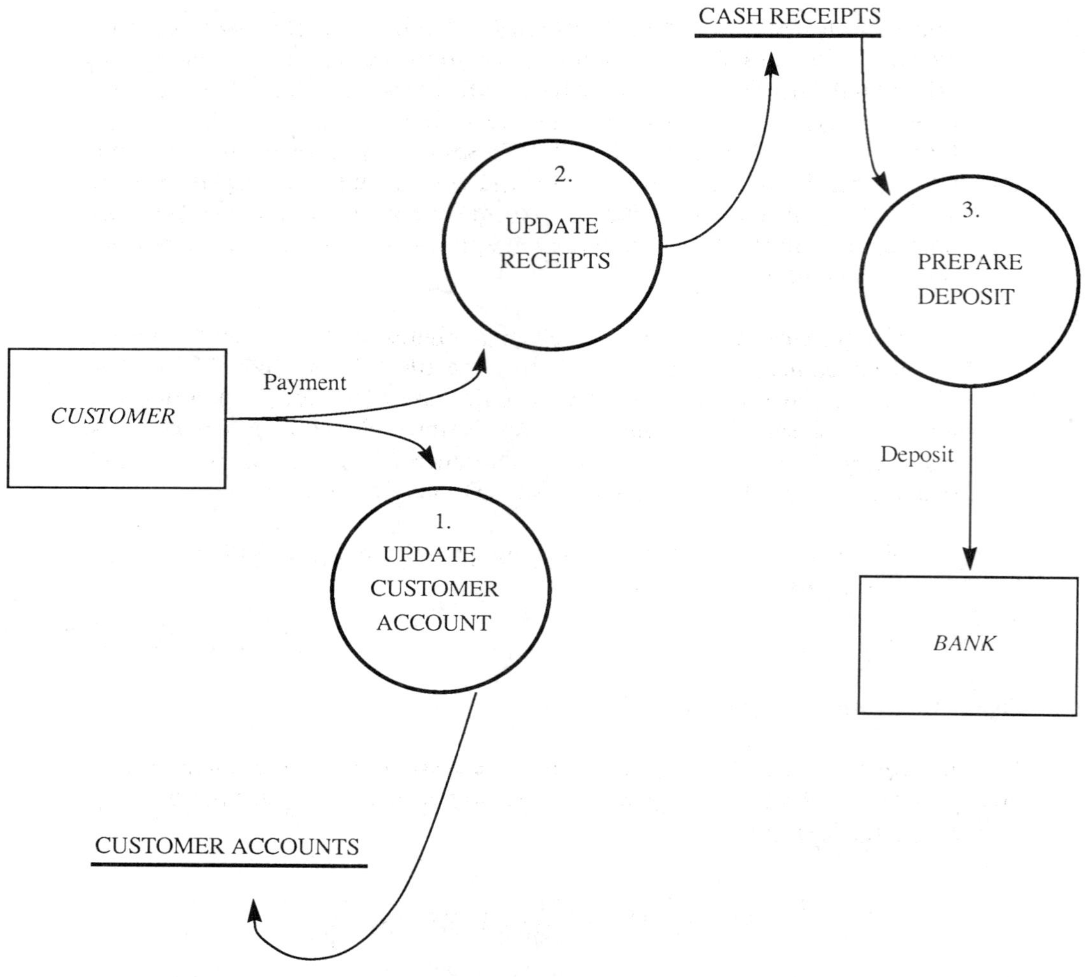

Figure 62: Excerpt of logical DFD

The excerpt in Figure 62 illustrates data file usage. The direction of arrows indicates whether data is flowing into or out of a file. Data flows into and out of files need not be named because the content of such a flow can be surmised by examining the data dictionary entries that contain the attributes of the file (entity). All other flows in a DFD should be named. Information contained in the file CASH RECEIPTS is needed for preparing the bank deposit, so the flow is from the file and to the process PREPARE DEPOSIT. A principle of data flows and files is that the *net* flow of data is indicated by the arrow. For example, in the process UPDATE CUSTOMER ACCOUNT shown in Figure 62, the file CUSTOMER ACCOUNT must be read first to find the proper account, so the flow will be *out* of the file and into the process. However, the purpose in reading the file is to output the proper data to the file, so the net flow is *to* the file.

The physical DFD excerpt shown in Figure 61 has a double-headed arrow indicating data flow to and from the file CASH RECEIPTS. This bidirectional arrow indicates data accessing as well as data updating, the two not necessarily occurring at the same times. Sometimes MARY adds data to the file (when processing a customer payment); at other times (when preparing the bank deposit), she extracts data from it.

Bidirectional data flows occur only to and from files. All other data flows in a DFD are unidirectional.

7.4 Conventions

To illustrate how data flow diagrams are developed, as well as the conventions and rules they follow, we begin with a case study, and then conclude by summarizing guidelines in point form.

FIRST CITY COMPUTERS, INC.

First City Computers, Inc. is a company that buys and sells specialized electronic components for all sizes and makes of computers. The company has experienced startling success since opening for business three years ago. Because of the considerable volume of activity the company is experiencing, management have decided that more formal procedures are now necessary. They wish to computerize the processing of customer orders and accounts, and inventory control operations, all of which are manual. (The maxim "those who automate others are themselves the least automated" applies to First City.)

Management feel that automation is necessary because of the long turnaround time for their accounts receivables: the average time between receiving an order from a customer and receipt of payment is 72 days. If orders, invoices and statements were processed more rapidly, payments would be remitted sooner. The company could apply these payments against their bank loan more quickly, and in doing so, reduce the amount of interest paid out. Management anticipates that improved control of inventory will also be a benefit of automation: inventory levels could be reduced. The purchase of inventory is financed by the company's bank loan.

An independent consulting firm, Second Software Consultants, has been hired to analyze the functions of order processing, accounts receivable and inventory control (including purchasing) and to apprise management of the benefits of automation. After several preliminary interviews with users, the senior systems analyst made the following list of observations concerning First City's operations:

ORDER PROCESSING

- Salespeople call on customers for orders. Orders are brought to Sam, who verifies information regarding the customer and the items ordered.
- For a new customer, Sam fills out a customer card with the person's name and address, and files it in the customer card file.
- For existing customers, he looks up their account balances in the customer card file, and if money is owing, he puts their orders into the order suspense file, where they are held until payment has been received.
- For a customer whose account is paid up to date, he checks the order against the inventory card file to verify that the parts ordered have the correct part numbers. He then gives the order to Sarah.
- Sam goes through the order suspense file each day, comparing it to the customer file to see if payments have been received. If so, he verifies the items ordered against the inventory, and gives the activated orders to Sarah.
- Sarah receives orders from Sam. She checks the quantities ordered against those quantities on hand recorded in the inventory file.
- If a customer order can be partially or completely filled, Sarah adjusts the quantities on hand accordingly on the card file, and enters the quantities shipped into the appropriate column on the order.

- The order is a two-part form. The copy (second part) is used as a picking slip (for locating the items in the warehouse) by warehouse staff.

- Accounts receivable (A/R) is responsible for recording payment amounts and invoice amounts on the customer file, and for sending invoices and statements to customers.

- Sarah sends the original order form to A/R, where the prices are calculated, and the dollar amount of the order is entered on the customer's card. An invoice is then mailed to the customer.

- For entire or partial orders that cannot be filled, Sarah prepares a back order and files it in the back-order file.

- Inventory control (I/C) is responsible for maintaining and ordering stock.

- At the end of each day, Brian, the inventory control clerk, updates the inventory stock cards with the receipts of the day.

- Each morning, Sarah goes through the back order file and compares the items back ordered to the corresponding on-hand inventory quantities. If a back order can be completely filled, she activates it by sending it to Sam for credit verification, the standard procedure for all orders. (Credit verification is performed for back orders because the status of a customer's account may have changed since the order was originally verified for credit.)

The physical data flow diagram for the order processing (O/P) function of First City is shown in Figure 63.

In physical data flow diagrams, the processes are often people or departmental names. All of the distinct activities performed by Sam are interior to the process named SAM, and we are concerned only with the flow of data to and from him.

The two functions *ACCOUNTS RECEIVABLE* and *INVENTORY CONTROL* are external to order processing, the area currently being investigated, and are shown as sinks. However, A/R and I/C interface with *ORDER PROCESSING* via the CUSTOMER ACCOUNTS and INVENTORY files, respectively.

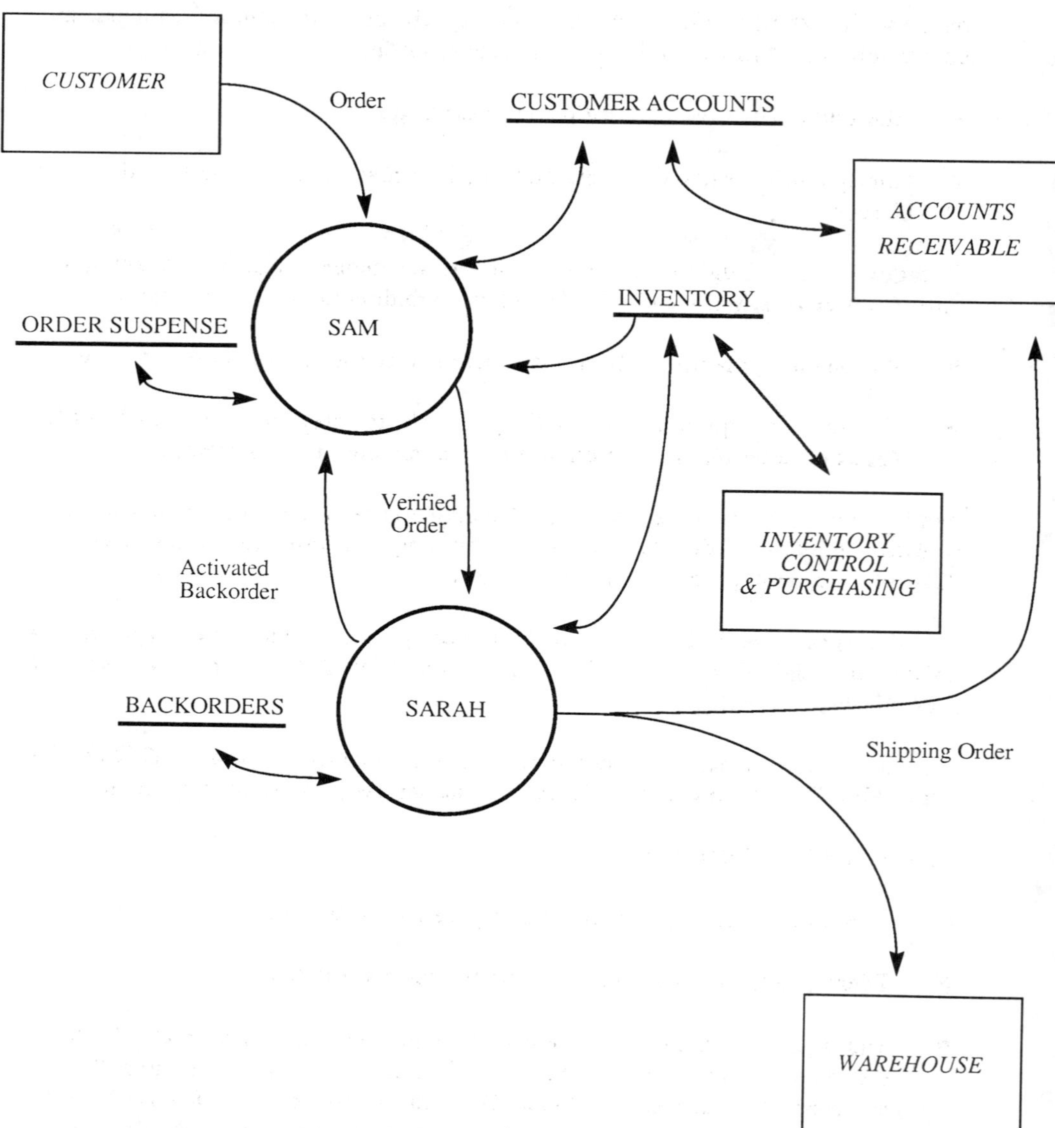

Figure 63: Physical DFD First City Computers ORDER PROCESSING

Notice the bidirectional arrows between SAM and CUSTOMER ACCOUNTS, SAM and ORDER SUSPENSE, and between SARAH and INVENTORY and SARAH and BACKORDERS. Three of the four are obvious, but the bidirectional arrow between SARAH and INVENTORY warrants an explanation. SARAH compares quantities ordered by customers with what is on hand in INVENTORY. If there is enough stock to supply the customer completely, the arrow is bidirectional, indicating that

- the quantities have been adjusted accordingly
- the quantity ordered can be shipped, information that is indicated on the Order.

If there is not enough inventory to fill the customer's order, then again the arrow between SARAH and INVENTORY is bidirectional in which case

- the on-hand quantity is reduced to zero, so the arrow is toward the file
- the on-hand quantity is used by Sarah in preparing the back order (back-ordered quantity = quantity ordered - quantity on hand).

Notice that there is an exception to the above case: when quantity on hand = 0, then the arrow is directed out of the file only. In order to present a correct diagram, the most general case must be shown.

We expand our discussion of First City physical data flow diagrams by examining *ACCOUNTS RECEIVABLE*, and *INVENTORY CONTROL and PURCHASING*.

The senior analyst interviewed users involved with *ACCOUNTS RECEIVABLE* at First City Computers and observed the following points:

ACCOUNTS RECEIVABLE

- A/R are responsible for maintaining customer accounts.
- There is one person, Anne, who performs this function.
- She opens the mail, and when a customer has sent a payment, she notes the payment amount and date on the customer's card, and updates the corresponding account balance. In addition to receipt of payments by mail, payments may be brought in by customers and by sales personnel.
- As payments come in throughout the day, she records each payment in the cash receipts journal.
- At the end of the day, she totals the journal entries, and deposits the receipts in the company's bank account.

- Anne receives the original customer order form for orders that have been shipped. These forms are used as the basis for calculating the customer's bill (invoice). She calculates the price for each item by looking up the current price in the company catalog and then calculating the total amount owing for each order.

- She sends invoices to customers, and for each customer account, updates the customer's card with the invoice amount, invoice number and date.

- Anne sends a copy of each invoice to the sales manager and files the remaining copy in the numerically ordered invoice file required by the federal government.

- On the last day of each month Anne

  - surveys all customer accounts and sends statements to customers with outstanding balances

  - totals the sales and receipts for the month, and gives these figures to management.

INVENTORY CONTROL and PURCHASING

The following observations concerning *INVENTORY CONTROL and PURCHASING* were made by the analyst:

- *INVENTORY CONTROL and PURCHASING* have responsibility for the procurement and maintenance of reasonable levels of inventory items.

- Brian is responsible for the day-to-day I/C and Purchasing functions. A file of inventory items is the main tool for controlling inventory.

- The file of inventory items is composed of individual stock cards, one for each item. Items are purchased when the on-hand quantities of stock fall below the reorder level indicated on the card. Quantity On Hand, Reorder Level, and Unit Cost are fields in the inventory card file. Brian updates these as required.

- Accounts payable (A/P) are responsible for paying suppliers from whom First City have purchased goods and services.

- Each Monday, the inventory file is searched, and for those items that need to be reordered, a purchase order is generated as follows:

- Brian locates the appropriate contract that forms part of the supplier's file.

- He prepares a purchase order, sending the original to the supplier, one copy to receiving, one copy to accounts payable, and filing one copy in the purchase order file.

- If there is no contract for the stock being reordered, he prepares a request for quotation, copies of which are sent to several prospective suppliers.

- Brian reviews the quotations submitted, chooses a supplier and files the quotation. The purchase order is prepared as above.

- Receipt of goods is acknowledged by warehouse employees who send a receiving slip to Brian.

- Brian matches the receiving slip with the corresponding purchase order. He updates the inventory file with the quantities indicated on the receiving slip and with any unit cost changes indicated on the purchase order. These two documents are filed in the purchase history file. The file is reviewed periodically by management.

- Suppliers send invoices to A/P, who forward them to Brian for validation. Brian validates the invoice, comparing it with the original purchase order and corresponding receiving slip, by signing his initials. He then returns the invoice to accounts payable.

- At the end of each month, Brian provides management with the total value of inventory on hand by calculating the sum of quantity on hand for each inventory item times its unit cost.

The physical DFD's for *ACCOUNTS RECEIVABLE* and *INVENTORY CONTROL and PURCHASING* are shown in Figures 64 and 65, respectively.

7.5 Developing a Set of DFD's

We are now in a position to discuss the derivation of the *logical* equivalents of Figures 63, 64 and 65, and to introduce the notion of systems decomposition via logical data flow diagrams, also referred to as structured analysis.

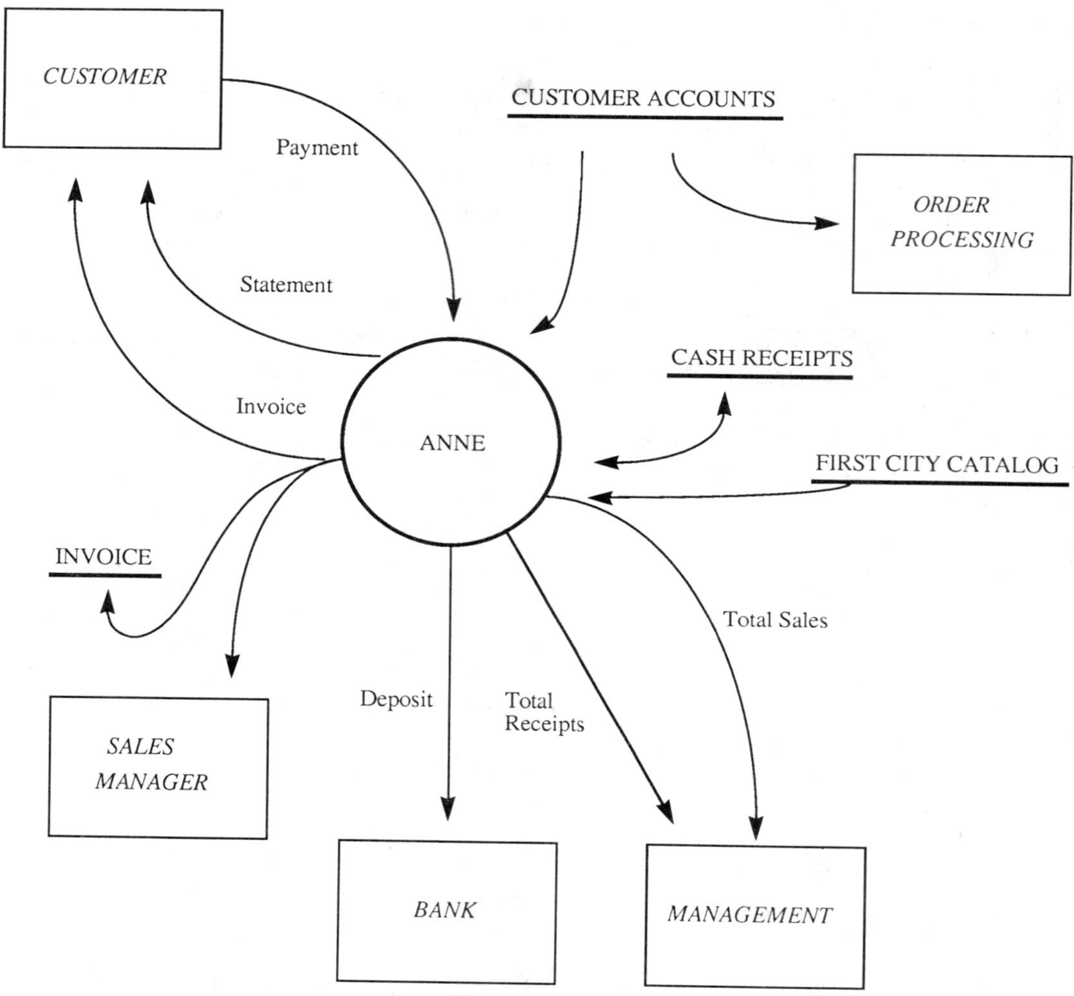

Figure 64: Physical DFD First City Computers ACCOUNTS RECEIVABLE

As mentioned previously, during the second phase of analysis, the analyst must abstract the logic underlying the physical DFD's produced during the first stage to produce the corresponding logical DFD's. We leave it to the reader to verify that Figures 66, 67 and 68 can represent the logical diagrams corre-

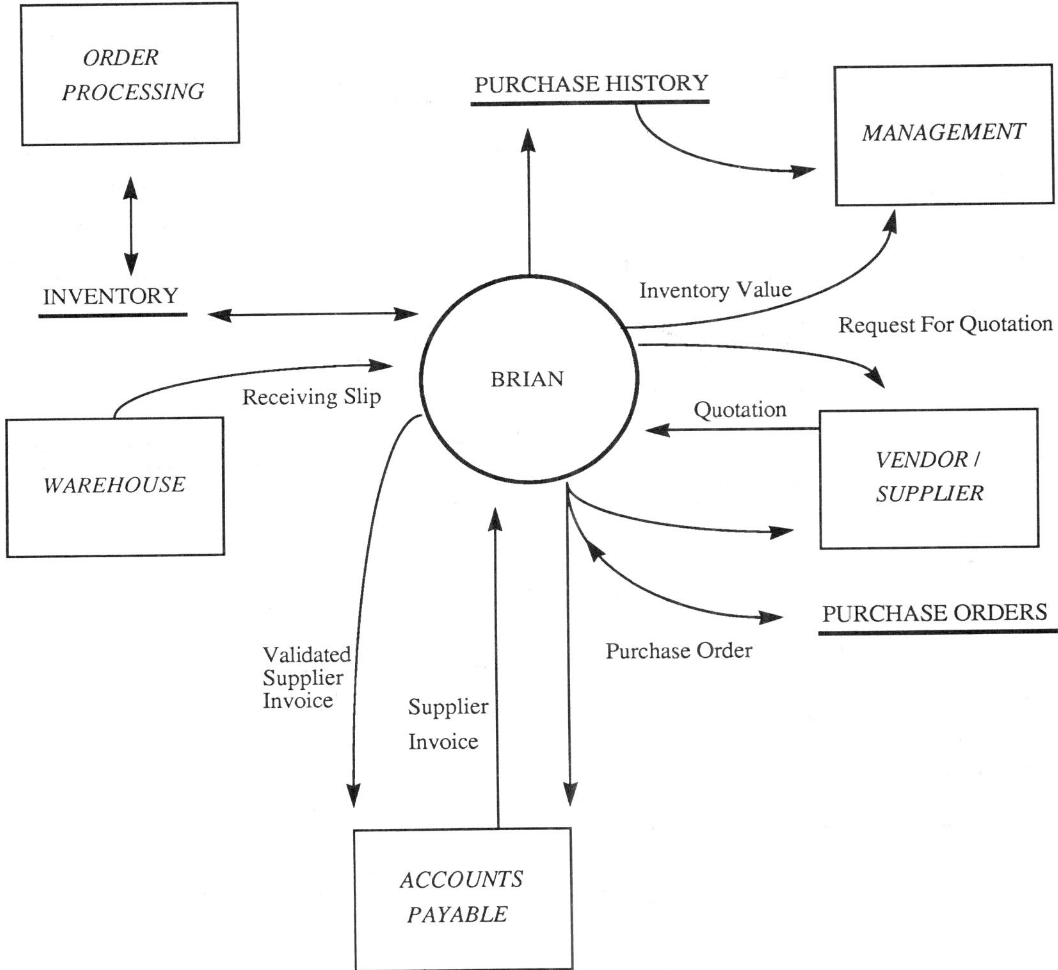

Figure 65: Physical DFD First City Computers INVENTORY CONTROL and PURCHASING

sponding to Figures 63, 64 and 65, respectively. (Figures 66, 67 and 68 reflect the authors' perceptions. There is not just one correct way to draw a data flow diagram, so your own version may be somewhat different. A marvellous aspect of DFD's is that there is more than one correct way to model any situation.)

When drawing logical DFD's, the following rule regarding files applies. Files are shown only when they are interfaces between or among DFD components. The file called FIRST CITY CATALOG, used in accounts receivable to prepare customer invoices, is used by one process only, and so is not shown. This file is *local* to the process PREPARE CUSTOMER INVOICE. As well, notice in Figures 66, 67 and 68 that some of the flows and sinks from the physical DFD's have been given more functional names.

Figures 66, 67 and 68 can be combined to form a single diagram (Figure 69), depicting an overview of the areas we have been investigating (O/P, A/R and I/C). Following DeMarco [6] we use the term *context diagram* for such a diagram. A context diagram can be defined as the top-level diagram in a set of DFD's. In a context diagram the area under study is encompassed by a single process ("the system"), and only net interfaces with other systems are represented. For First City Computers, the system we are studying consists of the functions ORDER PROCESSING, ACCOUNTS RECEIVABLE and INVENTORY CONTROL and PURCHASING.

The purpose of the context diagram is to give an overview of (define) the area under consideration. Several of the files shown in Figures 66, 67 and 69 do not appear in Figure 69. This is because they are internal or local to the system under consideration. (The rule discussed previously regarding files states that files are shown only when they are interfaces among DFD components. This rule applies to all the files that have disappeared.) The only file that interfaces with an external system, and is therefore shown, is PURCHASE HISTORY.

The diagram beneath the context diagram, showing the next level of detail, is called *Diagram 0*. Diagram 0 shows the division of a system into major functional areas or subsystems. Figure 70 shows Diagram 0 for the system we have been considering: the processes ORDER PROCESSING, ACCOUNTS RECEIVABLE, and INVENTORY CONTROL and PURCHASING. In Figure 70, each process has an identifying number.

We can now return to Figures 66, 67 and 68, and complete them by naming them Diagrams 1, 2 and 3, respectively. The processes of each diagram are also labeled. Figures 71, 72 and 73 are the resulting revisions.

System decomposition (or *partitioning* or *levelling* as it is also called) could continue further still. For example we could make process 1.1, VERIFY CUSTOMER ACCOUNT, into a diagram. Examining processes in increasing levels of detail is called *explosion.* If process 1.1 were exploded into a diagram, it would be labeled diagram 1.1, VERIFY CUSTOMER ACCOUNT. The processes within it would be identified as 1.1.1, 1.1.2, 1.1.3 and so on.

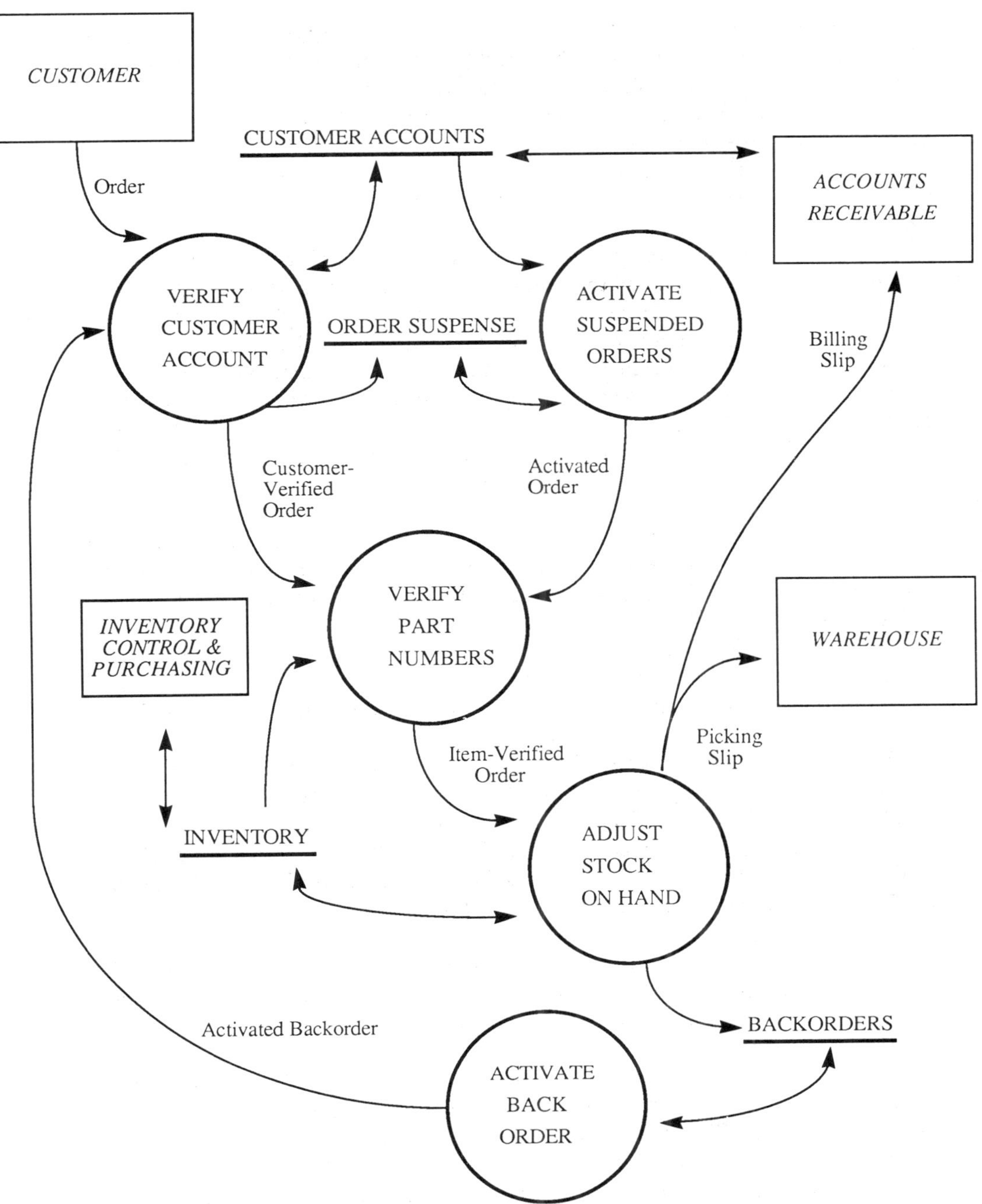

Figure 66: Logical DFD First City Computers ORDER PROCESSING

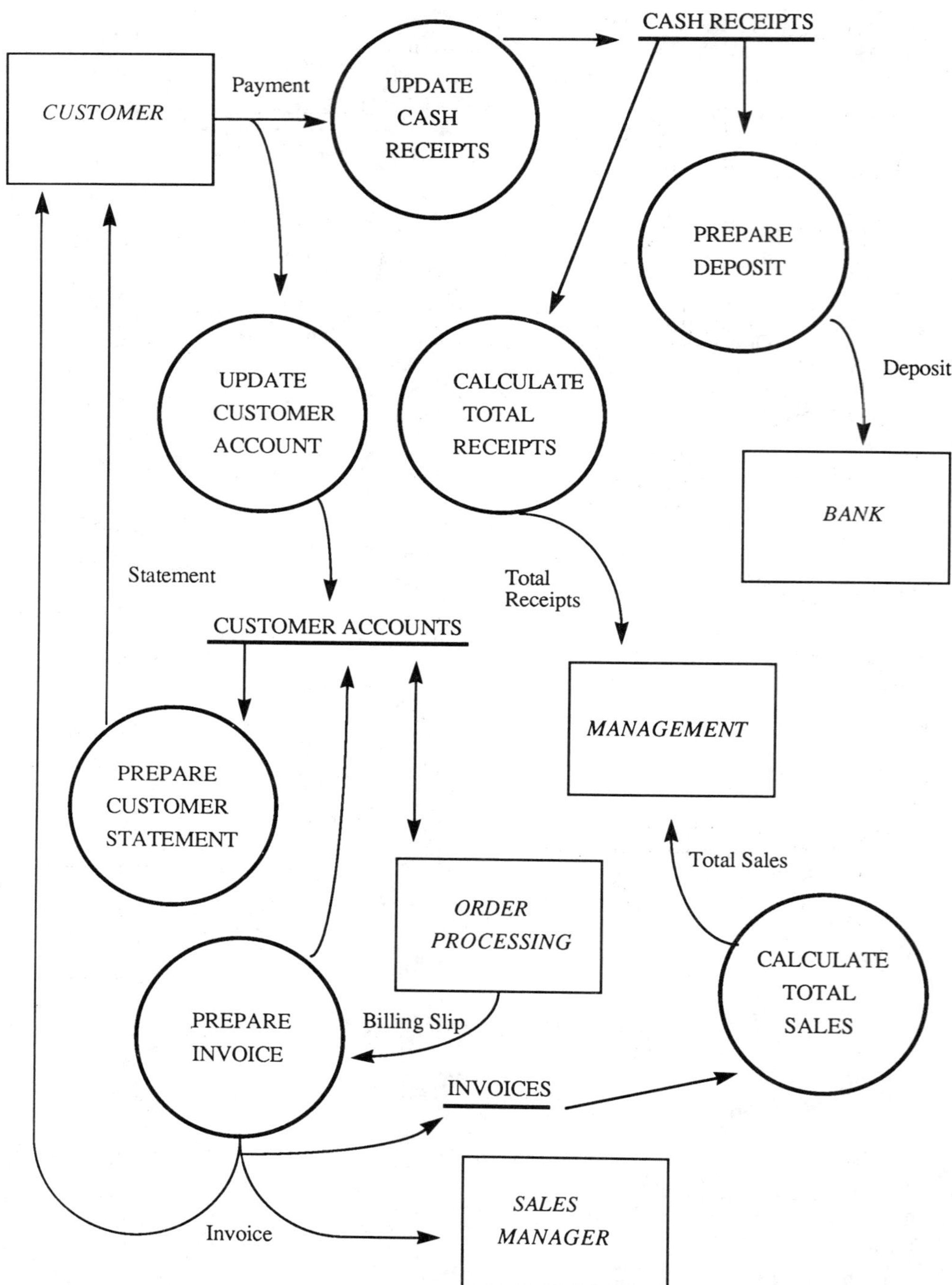

Figure 67: Logical DFD First City Computers ACCOUNTS RECEIVABLE

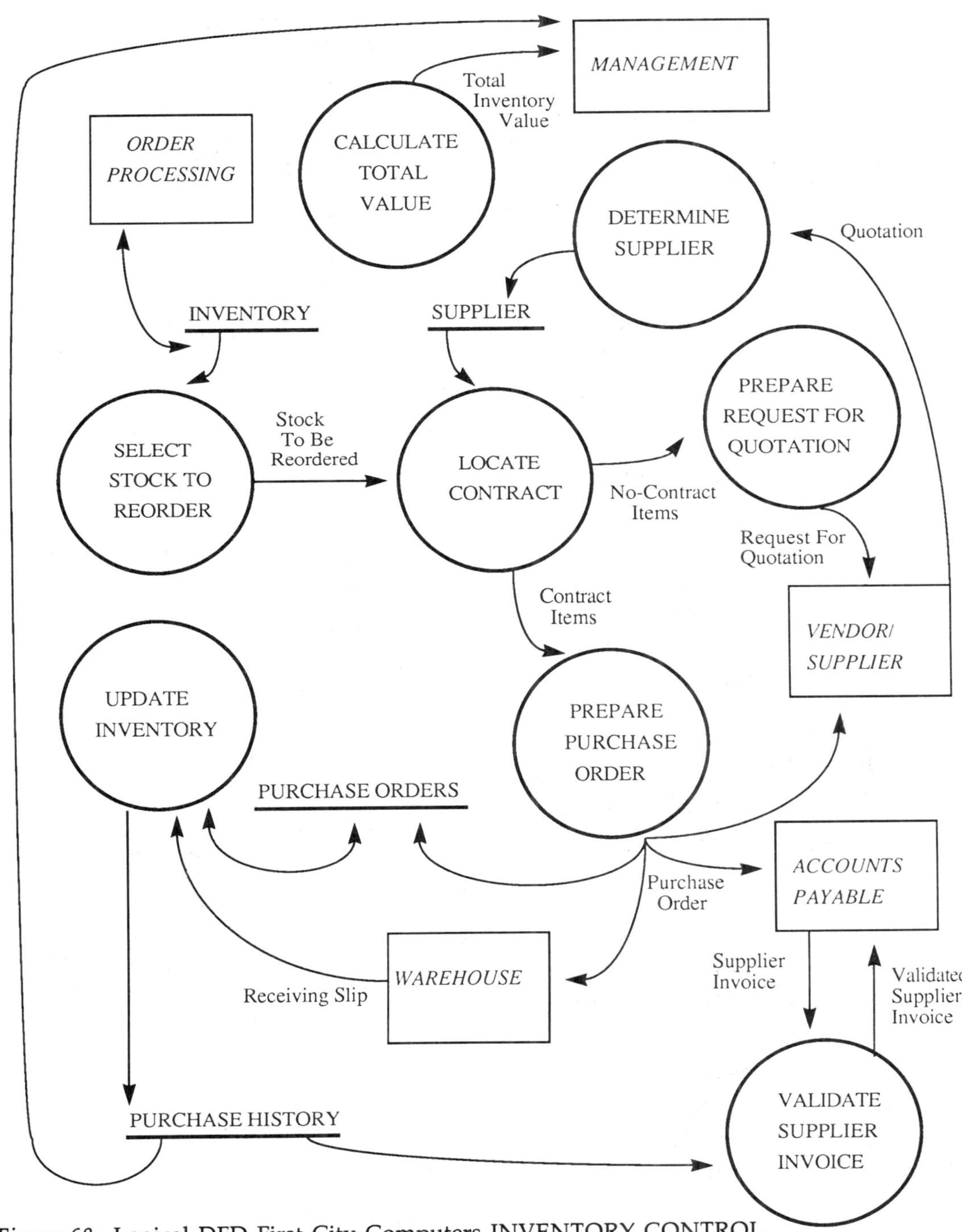

Figure 68: Logical DFD First City Computers INVENTORY CONTROL and PURCHASING

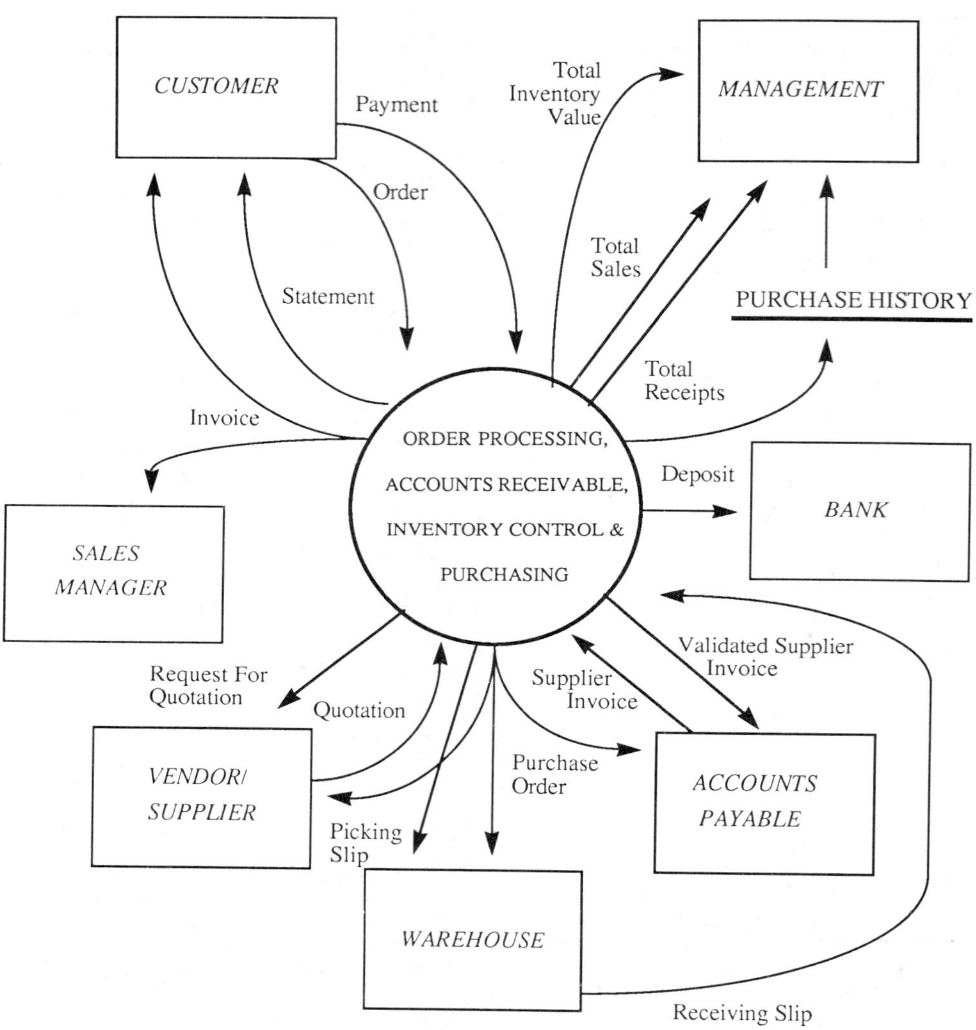

Figure 69: Context diagram First City Computers ORDER PROCESSING, ACCOUNTS RECEIVABLE, INVENTORY CONTROL and PURCHASING

In practice, how far should one continue to decompose a system? The bottom-level processes are the basis for future program modules, coded during the programming phase of development. The most maintainable program modules are the ones that perform exactly one function. With this in mind, system decomposition should continue until each process of a DFD performs one (and only one) function. If the process can be described by a single forceful verb and one definite object, as in "verify part number" or "activate back order," then the analyst has gone far enough. (Interiors of bottom-level processes will be discussed in Chapter 8.)

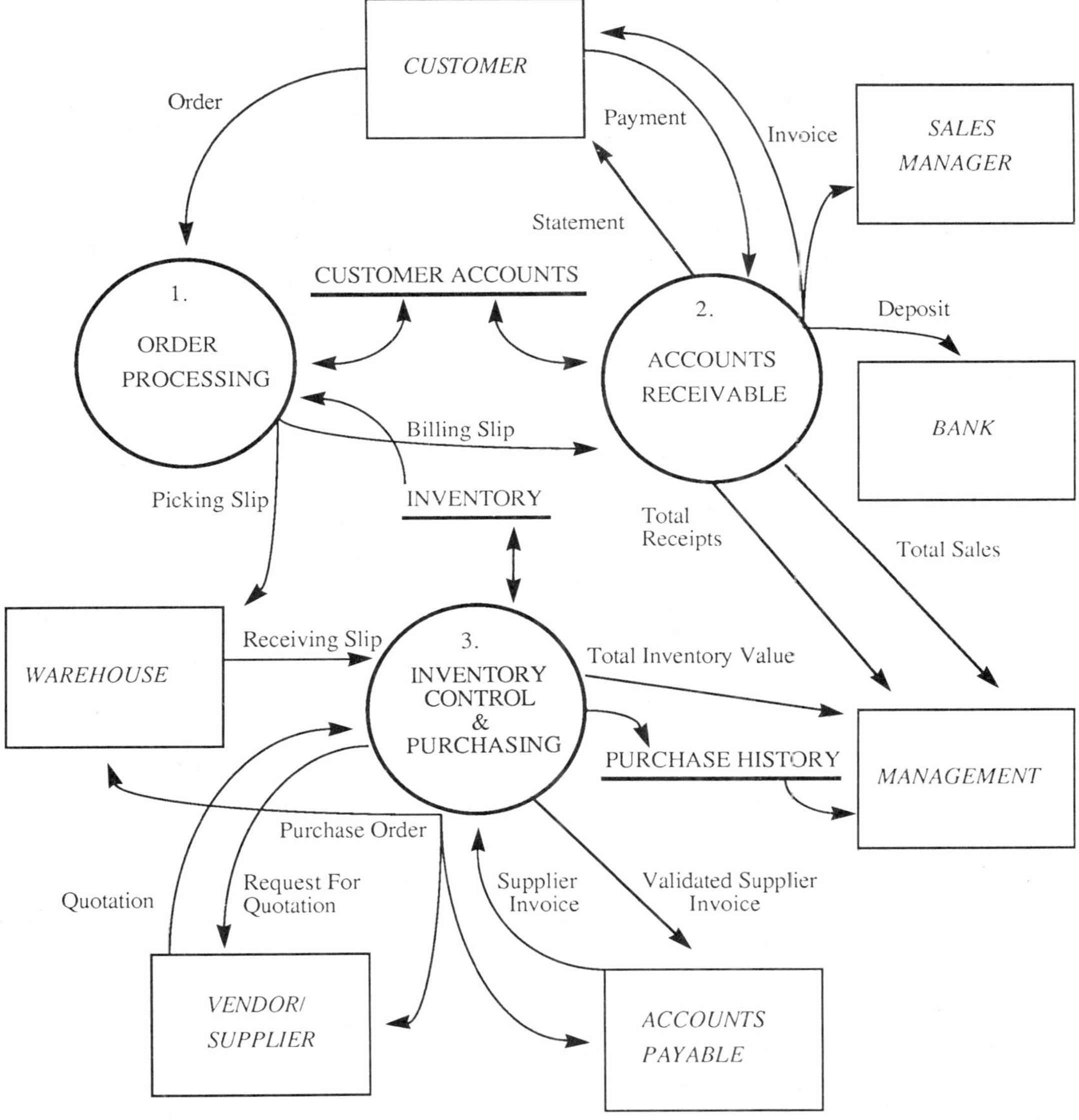

Figure 70: First City Computers Diagram 0

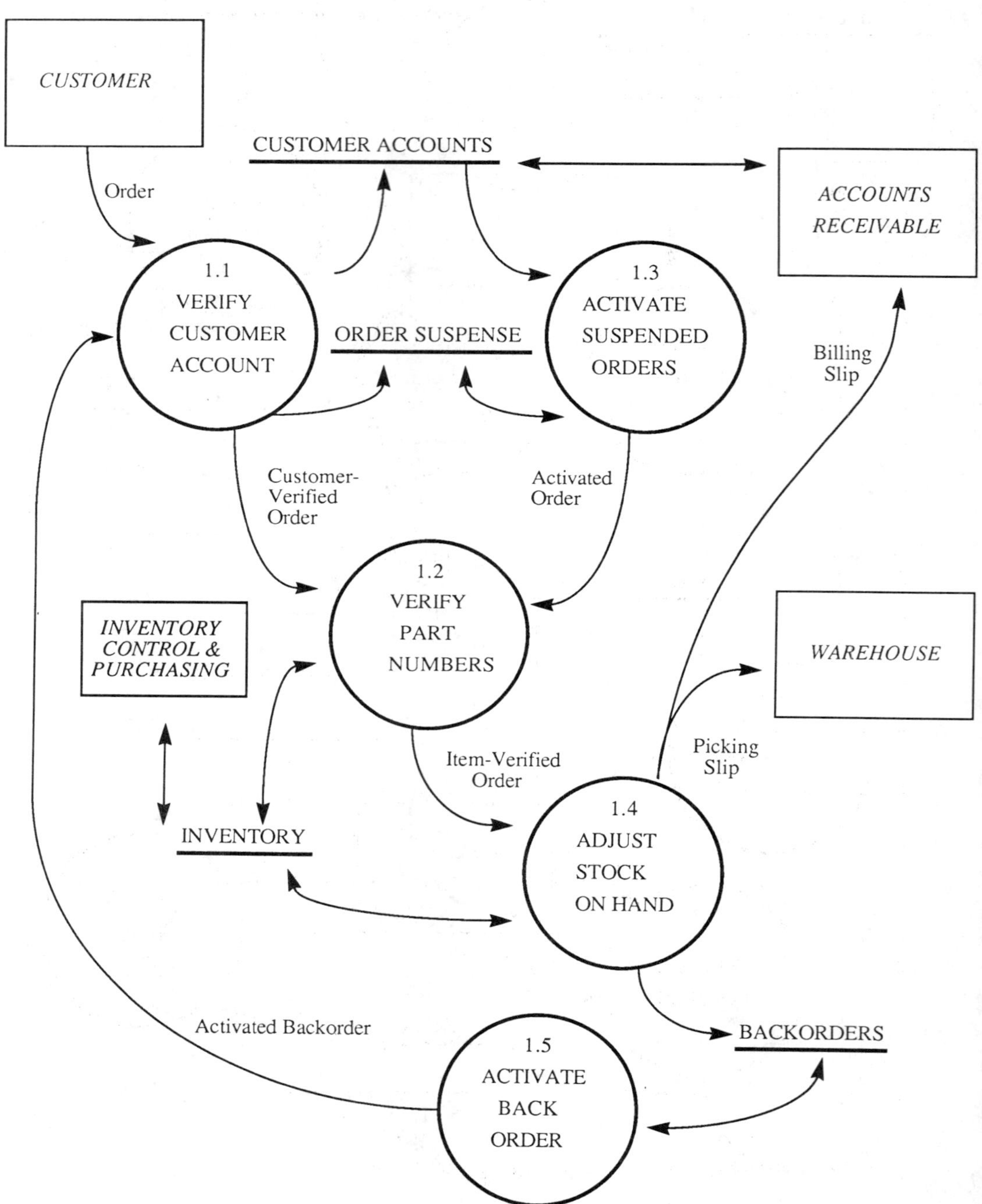

Figure 71: Logical DFD First City Computers Diagram 1

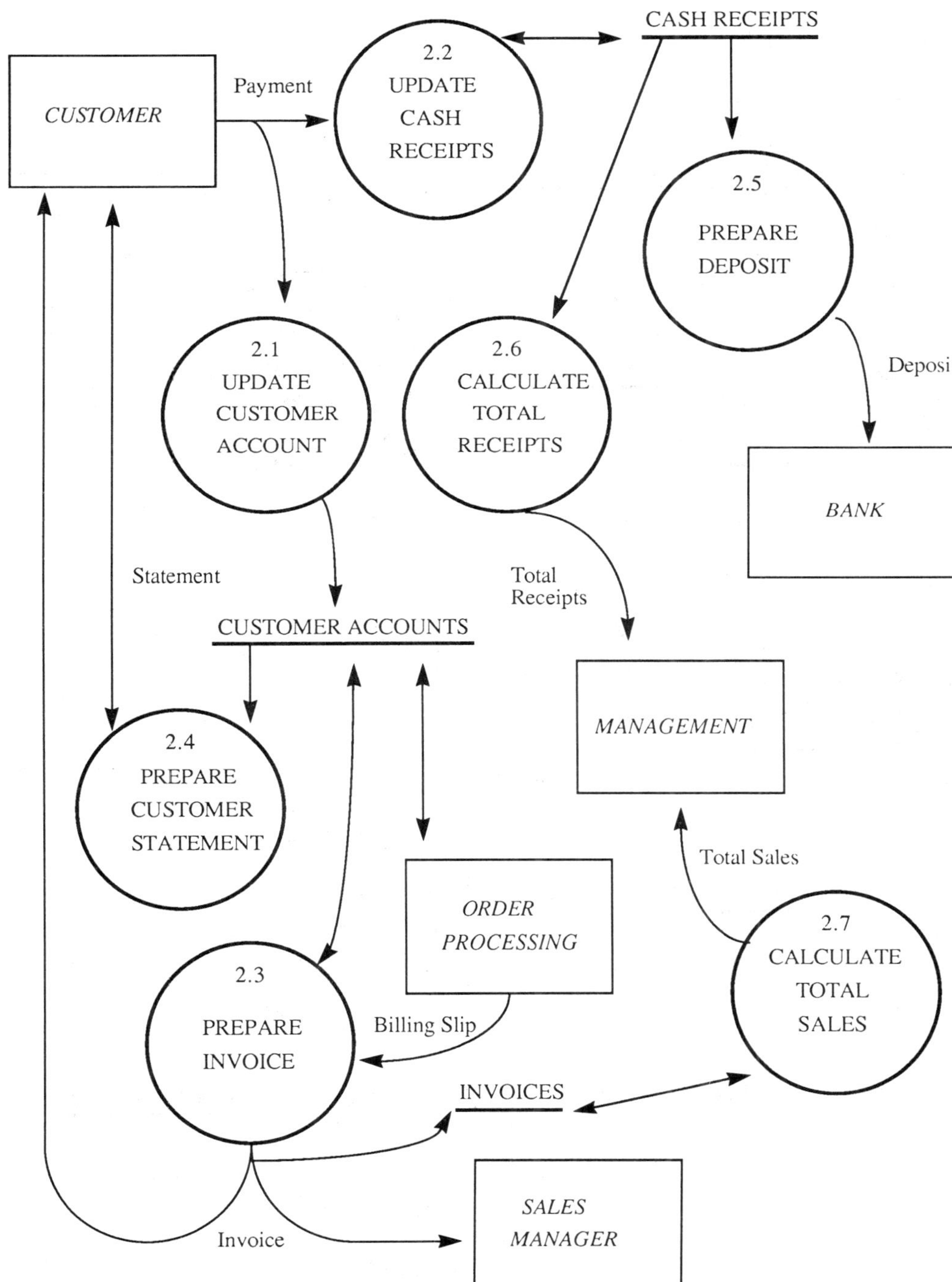

Figure 72: Logical DFD First City Computers Diagram 2

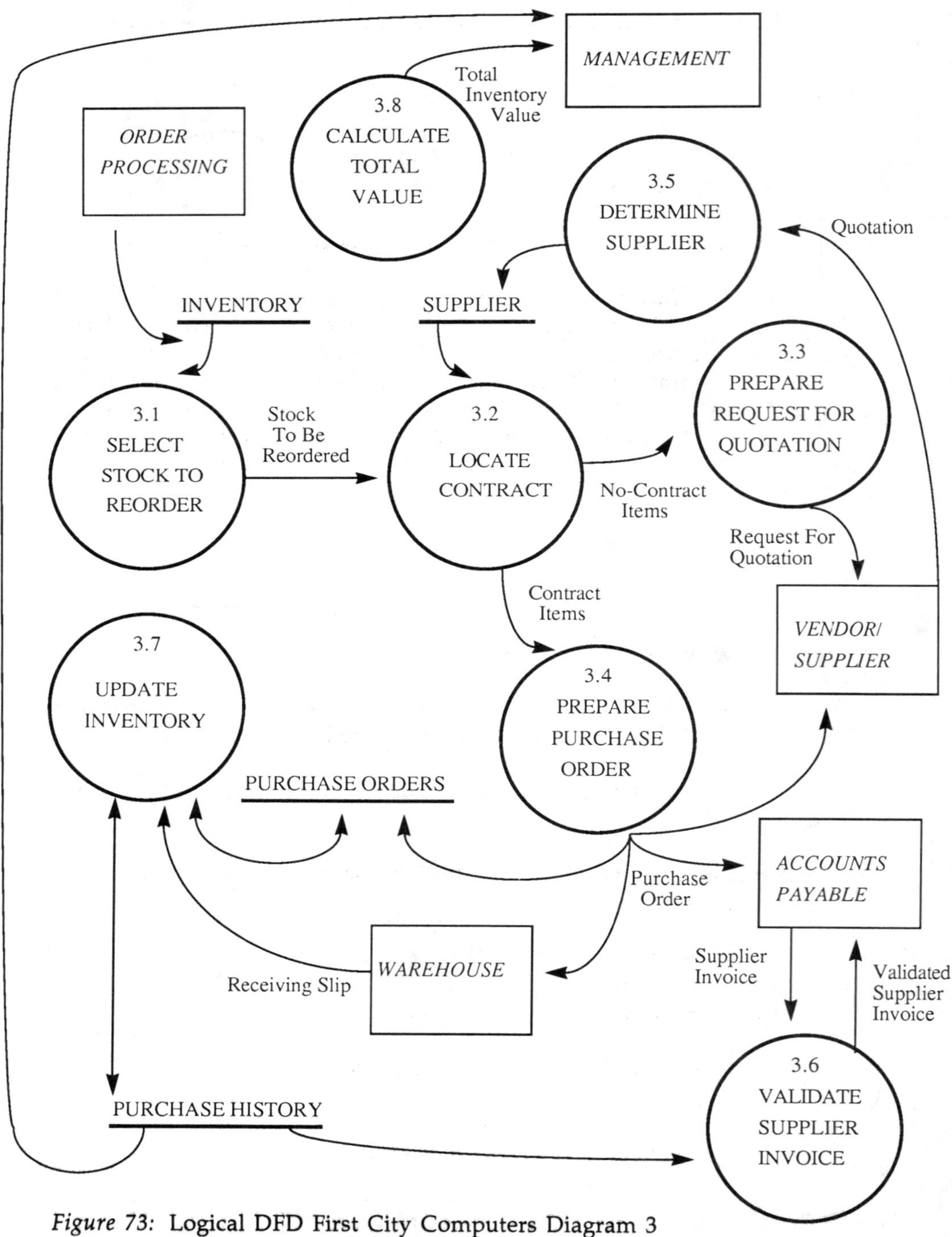

Figure 73: Logical DFD First City Computers Diagram 3

For this simplified case study, the set of logical DFD's has three levels:

- context,
- diagram 0
- diagrams 1, 2 and 3.

In a real application, several more levels would probably be required.

We make one final point regarding the processes on a DFD. Processes generally fall into one of two categories: they either transform (change) the data in some way, as in process 2.7 of Figure 71, CALCULATE MONTHLY SALES, or they filter (separate or direct) data as in process 1.1 of Figure 70, VERIFY CUSTOMER ACCOUNT, in which the incoming orders are separated into two types, customer-verified and suspended. Although this distinction may not be important during analysis, it is important during the design phase of development.

7.5.1 Developing a Set of DFD's: Summary

Rules, conventions and guidelines for drawing DFD's have surfaced in discussions. Aspects of developing data flow diagrams are listed below.

- DFD's may be physical or logical, and may be depicted in various levels of detail.
- External systems interface with the one under study through data flows and data files.
- A file is a place where data resides. It is shown first at the level of detail at which it interfaces with other components. (Otherwise it is interior to one of the processes, and so is not shown.)
- In a complete set of logical DFD's, each process has an identifying number.
- Only net flows to and from files are shown.

- For correctness, the most general case is depicted.

- Two independent processes that use the same data flow(s) are shown as being parallel (processes 2.1 and 2.2 in Figure 72).

- Two processes, one of which depends on the data output from the other are shown as being serial (processes 1.2 and 1.4 in Figure 71).

- Flows to and from files are the only ones that need not be named. Flows to and from files are the only ones that may be bidirectional.

- The most detailed level in a set of DFD's is the level at which the processes can be described by one forceful verb and one definite object.

- The product of the first phase of analysis is a leveled set of data flow diagrams describing the present system. These diagrams are called "Current Physical DFD's" in Figure 58.

- The product of the second phase of analysis is a leveled set of logical data flow diagrams describing the present system. These diagrams are called "Current Logical DFD's" in Figure 58.

- A set of leveled logical data flow diagrams of the system conceived by the analyst designed to meet the user's business needs is one of the products of the third phase of analysis. These diagrams are called "New Logical DFD's" in Figure 58.

Data flow diagrams depict the steady state of a system. As with any modeling tool, they protect us from a lot of detail. During the last phase of analysis, after management have chosen an automation strategy, essential details showing the non-steady state of the system are added (paths, logic, etc. for error detection) to provide a complete picture for the design phase of development.

For clarification, we make a final comment concerning developing a leveled set of DFD's. A set of leveled data flow diagrams model a system in all levels of detail, from an overview (context diagram) to the most detailed level (bottom-level DFD's). In practising analysis, an analyst seldom commences with an overview, or with the bottom level, but begins somewhere in between. Through iteratively developing and refining the model, the analyst's understanding of the system increases, and ultimately results in a leveled set of DFD's.

7.6 Exercises

1. Data flow diagrams of the solution conceived by an analyst are a major part of the functional specification. In this chapter, considerations regarding redundancy among the diagrams in a leveled set of DFD's were not discussed. (Examples of redundancy can be found in the set of diagrams of Figures 69 through 73. Figure 69, the context diagram for the order processing, accounts receivable, and inventory control and purchasing functions at First City Computers, shows data flows going to and from sources and sinks. These flows and sinks are repeated in the lower-level diagrams.)

   a. Why should redundancy be a consideration? What undesirable consequences could result from a leveled set of DFD's of the new system containing redundancies?

   b. Develop a guideline for drawing DFD's that addresses the redundancy issue.

2. Can a process invent data, or must there always be a data flow input to a process, transformation by the process and then output from the process? Illustrate your answer with examples.

3. Suppose you are analyzing a system and documenting your analysis using data flow diagrams. One of your diagrams contains a file with two arrows directed toward the file and none directed away from the file. What conclusions can you draw?

4. Except for diverging flows, why should data flow names be unique?

5. What is the role of the data dictionary during analysis?

6. Develop a leveled set of physical data flow diagrams for the following description of commercial airline travel, from ticket purchase to take-off.

   A prospective passenger phones a reservation office to purchase a ticket and reserve a place on a particular flight. The reservation agent is given the following information: city pair, date of departure, time of departure, fare class (first, economy or discount) and any special service requirements (wheelchair, meals). Passenger name, address, phone number and credit card number are additional required items. During the phone call, as information is being recorded, the agent accesses the file FLIGHT INVENTORY to see if a seat on the required date at the desired price is available. If not, other flights and classes are checked until one meeting the passenger's requirement is found. This being accomplished, the FLIGHT INVENTORY file is updated. As well, the file CUSTOMER DATA is updated with all of the passenger information including the flight number, date, fare class and selected seat. A record of the ticket sale is sent to the financial database. A few hours after completion of these procedures, a ticket is issued and mailed to the passenger's home. An update of the CUSTOMER DATA file is made to this effect.

On the day of departure, passengers arrive at the appropriate counter for check in. Their names and ticket numbers are entered into the reservation system that accesses the CUSTOMER DATA file and displays passenger data. Simultaneously, two other files are accessed, the SEAT AVAIL file and BOARDING AUTHORITY file. The process here is as follows.

If a passenger is confirmed for the desired flight (as described above), the files will already contain the required data (name, flight number, etc.). However, the check in process must accommodate passengers who arrive with a ticket but with *no* reservation. For confirmed passengers arriving with a ticket and without a reservation, the available passenger data contains everything but a flight number, time and assigned seat. For confirmed passengers arriving with a ticket and with a reservation, the SEAT AVAIL file contains the confirmed passenger's name assigned to a particular seat. (This was performed automatically when the CUSTOMER DATA file was updated upon reservation.) Unassigned seats (if any) are indicated by the lack of an assigned name. If a passenger is not confirmed and there is a seat available, the seat is assigned immediately and the passenger treated as confirmed. If there is no seat available, the passenger is registered as stand-by and the file STAND BY is updated. Upon check in of confirmed and assigned, and unconfirmed passengers (i.e., non-standby), the available seat file is updated, showing who has arrived. Also at this time, the CUSTOMER DATA file is read and a boarding pass is issued showing any special requirements (e.g., special meals, wheelchairs) and given to the passenger. One further aspect is that on a continuing basis, the available seat file is updated whenever a reservation is cancelled.

The BOARDING AUTHORITY file contains information such as what gate the flight departs from and what time boarding is to start. This information is given to the passenger.

Check in requires that baggage be processed as follows. The total weight of baggage is measured, and if in excess of 80 lb, extra costs are assessed. These costs are computed by accessing the BAG TARIFF file for that flight. All baggage is tagged and the tag numbers are entered into the CUSTOMER DATA file.

The check-in procedure has now been completed, and the passenger may proceed to departure gate. All confirmed passengers are seated on the aircraft five minutes before departure. Prior to this, boarding agents have collected boarding pass stubs and ticket coupons. Unclaimed seats are assigned to stand-by passengers, if any, by the boarding agent. Stand-by passengers are accommodated in the order in which they registered. As with confirmed passengers, a boarding pass is issued and a baggage stub and ticket coupon collected.

7. Develop a leveled set of logical DFD's of the payroll department described below.

Each Monday, Fred Jones receives time cards from various departments of the company. From these he creates a weekly time sheet by entering the date, each employee's name, employee number, regular hours worked and overtime hours worked. Unreadable or incorrectly filled in cards are corrected by phoning the appropriate supervisor for the correct information.

The weekly time sheet is next processed by Joan Fredericks who prepares pay slips for each employee. These contain the date, employee name, employee number and hours worked. She then calculates and enters the following: gross pay, federal income tax deduction, city tax deduction, charitable organization contributions and, of course, net pay. Gross pay is computed by referring to a pay-rate file organized by employee number. Taxes are computed by referring to files named FEDTAXINFO and CITYTAXINFO. Charitable donations are referenced in a file called CHARITY. Completed pay slips are forwarded to Fred Jones.

Fred now calculates company contributions for each employee to the government pension plan. Both the employee's deductions and company's contributions are entered in the files FEDTAXREMIT, CITYTAXREMIT, PENSIONREMIT, CHARITYREMIT. These files can be accessed by date or by employee name.

The pay slips are next sent to Cindy Rella who makes up individual employee pay checks from the net pay figure on the pay slips. Both the the pay checks and pay slips are sent to the departments from which they originated for distribution to employees. When pay checks are finished, Cindy summarizes and totals each of the remittance files for the week and sends a check and summary to the appropriate agencies. She also prepares a weekly payroll summary for the accounting department, which contains all information with respect to remittances and payroll.

8. An analysis of the Engineering Firm See-Saw Associates is currently underway. They wish to determine if there is a more effective way to administer their office procedures. A description of current procedures follows.

   Customers solicit bids on contracts, and grant or deny See-Saw's bid by comparing with other bids received. Sharon, the office manager, coordinates this activity. The client will send Sharon the project information, which includes drawings of the proposed job. She then turns these drawings over to the architects, who make detailed drawings of the job, along with cost estimates. The project manager, Stan, uses the drawings and the estimates to develop the bid for the contract. In order to accomplish this, he also reviews the suppliers and sub-contractor's files to determine material and labor costs. In addition to the bid, a contract is also prepared. The contract is reviewed by the firm's lawyers, and is then mailed to the client along with the bid. A copy of the bid and the contract is filed in the MAIN CONTRACT file.

   If either the client or the firm ends negotiations, the bid and the contract are marked void and filed in the MAIN CONTRACT file.

   If the contract is granted to See-Saw, then a summary card is begun for that project. Throughout the life of the job, Sharon's supervisor will be updating the summary card for each contract with information pertaining to time spent and materials used.

   At the end of each month, Sharon produces an item summary report which summarizes all of the items on the summary cards for that project. Based on this report, an invoice is prepared for the client.

   a. Draw a leveled set of physical data flow diagrams for the scenario described above.

b. Using your set of physical DFD's as the basis, derive a leveled set of logical DFD's for See-Saw associates.

9. Draw a leveled set of logical data flow diagrams for exercise 4b in Chapter 3. (These DFD's will complement the data model developed in 4a.)

10. Draw a leveled set of physical data flow diagrams that model the registration procedure at your college or university.

7.7 Bibliography

1. Biggs, Charles L., Birks, Evan G. and Atkins, William. *Managing the Systems Development Process*. Englewood Cliffs, New Jersey. Touche Ross and Company, Prentice-Hall, 1980.

2. Brown, David B. and Herbanek, Jeffrey A. *Systems Analysis for Applications Software Design*. Oakland, California. Holden-Day, Inc., 1984

3. Cash, James I., Jr., McFarlan, F. Warren and McKenney, James L. *Corporate Information Systems Management: Text and Cases*. Homewood, Illinois. Richard D. Irwin, 1983.

4. Dampney, C.N.G. et al. *Data (Information) Analysis of Business Systems - Informal Notes*. Waterloo, Ontario. University of Waterloo, 1987.

5. Davis, G.B. *Management Information Systems - Conceptual Foundations, Structure and Development*. New York, New York. McGraw-Hill, 1983.

6. DeMarco, Tom. *Structured Analysis and System Specification*. Englewood Cliffs, New Jersey. Prentice-Hall, 1979.

7. Gane, Chris and Sarson, Trish. *Structured Systems Analysis: Tools and Techniques*. Englewood Cliffs, New Jersey. Prentice-Hall, 1979.

8. Gildersleeve, Thomas R. *Successful Data Processing System Analysis, Second Edition*. Englewood Cliffs, New Jersey. Prentice-Hall, 1985.

9. Hodge, Bartow and Clements, James. *Business Systems Analysis*. Englewood Cliffs, New Jersey. Reston Publishing, 1986.

10. Jackson, M.A. *Principles of Program Design*. New York, New York. Academic Press, 1975.

11. Jackson, M.A. *System Development*. Englewood Cliffs, New Jersey. Prentice-Hall, 1983.

12. Parkin, A. *Systems Analysis*. London, U.K. Edward Arnold, 1980.

13. Parkin, A. *Systems Management*. London, U.K. Edward Arnold, 1980.

14. Pressman, Roger S. *Software Engineering: A Practitioner's Approach*. New York, New York. McGraw-Hill, 1982.

15. Senn, James A. *Analysis and Design of Information Systems*. New York, New York. McGraw-Hill, 1984.

16. Warnier, Jean Dominique. *Logical Construction of Programs*. New York, New York. Van Nostrand Reinhold, 1974.

17. Weinberg, Gerald M. *Rethinking Systems Analysis and Design*. Boston, Massachusetts. Little, Brown and Company, 1982.

18. Yourdon, Ed and Constantine, Larry L. *Structured Design*. New York, New York. Yourdon Incorporated, 1975.

Chapter 8

PROCESS MODELING

A software system is a model of some aspect of an organization, and a software system is comprised of the dual components, data and process (function). To be useful, the software system must be accurate enough to substitute for the aspect being modeled. For example, an inaccurate automated inventory control system cannot substitute for an accurate manual inventory control system. During analysis, the systems analyst gains an understanding of the business environment and business needs in sufficient detail to accomplish the objective of analysis: the derivation of the conceptual model of the software. Later phases of development will translate the conceptual model to a software system.

The previous chapter focused on data flow diagrams as a systems modeling tool, a tool that depicts the flow of data within a system and the functions performed by the system. Although the DFD's of the solution system conceived by the analyst depict the data and functions of a system, the DFD's do not **define** them. Data and function definitions are found in the data dictionary. Discussions in this chapter concern defining those functions of a system that evolve into computer programs.

One of the tasks performed by an analyst is to specify in detail the logic of the processes performed by the conceived system. (The flows among processes and files become the inputs and outputs to the programs.) Three different tools used for describing process logic are introduced in this chapter. These tools are structured English, decision tables and decision trees.

The three tools we present should allow the analyst to describe each process in a precise, concise way, clear enough that a program may be coded from the description. At the same time, we want the description to be nontechnical enough to be understood by the user. Normally only one of the three tools is used to describe any particular process. As well, different systems groups may prefer that only one tool be used throughout a specific project, or for all software development projects.

8.1 Structured English

8.1.1 Purpose/Use

The purpose of *structured* or *tight* English is to describe the logic of a process in an unambiguous yet easily read way. We can draw on the ideas from the data dictionary (Chapter 5) and from the discipline of structured programming to use the constructs of sequence, selection and repetition. Structured English is much like pseudocode, a term also associated with structured programming, but it is user friendly and less technical. We use it when the logic of a process is relatively uncomplicated.

8.1.2 Advantages

Consider the following structured English statement:

```
IF a student has completed 48 semester courses
     IF a student has completed the required 30 math courses
          THEN
               the student has earned a Bachelor of Mathematics degree
     OTHERWISE
               the student has earned a Bachelor's degree.
```

It states in a concise way what is required for a Bachelor of Mathematics degree. Structured English is a way of communicating the functioning of the processes to both the user and to systems personnel. Notice that it is understandable with little or no explanation and that it is also amenable to programming. Thus it bridges the gap between English and computer programming languages. Some of the words appear in caps. Such words and phrases are reserved for formulating logic. Phrases following a particular logic path are indented.

8.1.3 Where It Fits

The process logic is described during the third phase of analysis, modeling a solution system. After management have selected a system to be developed (phase 5 of analysis), some processes may have to be revised, some new ones added or some deleted. Structured English is used in phases 3 and 5 of analysis, as well as the other methods, decision tables and decision trees, which will be discussed later.

8.1.4 Components, Conventions, Rules

More than twenty years ago two Italian mathematicians proved that any computable logic is expressible using the constructs of sequence, selection and repetition. This proof lead to the discipline of structured programming, a discipline practised today in most systems groups.

Program modules are coded from the analyst's specifications. Specifying process logic during analysis, using the same constructs as the programs will use, allows the programmer to translate easily from specification to programming language. We briefly describe the three constructs.

- sequence - action(s) are to be performed sequentially, without interruption
- selection - one course of action among several is to be performed
- repetition - a set of actions is to be repeated within some limit

In our discussion of the components, conventions and rules of structured English, we will consider only these three constructs. As well, we require each process to have a single entry point and a single exit. Vocabulary will be limited to data dictionary terms whenever possible, and precise English otherwise. Sentences will be declarative. The First City Computers case study from Chapter 6 is used as a vehicle for discussion.

First City Computers

Refer to the Accounts Receivable function and its corresponding data flow diagram, Figure 72 (Diagram 2) in Chapter 7. Process 2.5 of Diagram 2 is titled Prepare Deposit. Suppose that after several interviews with users, the analyst derived the following description regarding preparation of the bank deposit.

At about 3:30 P.M. each day the bank deposit is prepared. We look in the cash receipts journal, where the payments are entered as they are received throughout the day, either by mail, by salesmen dropping them off, or by customers bringing them in. Currency and checks are kept in the cash box. All the receipts for the day are tallied; separate subtotals are kept for cash and for checks. The subtotals and total, a listing of each check amount, and an itemized listing of the denominations of the bills are entered into the deposit book supplied by the bank. The subtotals and total are also noted in the journal. The checks, cash, and deposit book are taken to the bank, where the amounts are verified and the book is stamped.

These procedures could be translated into structured English as follows:

```
FOR EACH of today's entries in the cash receipts journal,
    IF the payment is a check,
            denote the check amount in the deposit book,
            add the amount to the cumulative check total
    OTHERWISE
            add the cash amount to the cumulative cash total,
            add the number of 1's, etc., to each denominational total.

Enter the check and cash subtotals into the deposit book
and cash receipts journal.
Total the two amounts.
Enter the totals into the deposit book and cash receipts journal.
Place the cash and checks inside the deposit book.
```

This process has a single entry point, the first sentence, and a single exit, the last sentence. Notice that the process does not end with a statement of the deposit being taken to the bank. Only the *interior* of process 2.5 is described; the deposit going to the bank is described in the DFD itself (Figure 72), and it would be redundant to describe it here.

The previous example illustrates the use of all three constructs. A general framework for the selection construct is

```
IF  < condition >
        action to be performed if condition is true
            .
            .
            .
OTHERWISE
        action to be performed if condition is false
            .
            .
            .
```

An alternative form of this construct can be used in cases where several possibilities occur in one situation, but only one of which applies. For example, suppose the following narrative describes the customer discount policy at First City Computers.

Customer discounts are determined at the beginning of each calendar year, using the payment record of the previous year's purchases as the basis. For purchases in excess of $50,000, the discount is 20 percent, so long as the customer's payments have been prompt; otherwise the discount is 15 percent. For customers purchasing between $10,000 and $50,000 and making prompt payments, the discount is 10 percent; otherwise the discount is 5 percent. Others receive no discount.

One way to write this is shown in Figure 74. Note that we will have to clarify which case applies when the purchases are exactly $50,000 or $10,000. Also prompt and delinquent payment histories must be explained. Of course the analyst must take care that the process does not resemble programming language code or pseudocode too closely.

The framework for the repetition construct can be derived from the example illustrating deposit preparation at First City Computers on page 220.

```
<Condition for continued repetition>
        actions to be repeated
            .
            .
            .
```

Choose the applicable case:

CASE 1: previous year's purchases > $50,000 and
prompt payment history

discount is 20%.

CASE 2: previous year's purchases > $50,000 and
delinquent payment history

discount is 15%.

CASE 3: previous year's purchases between $10,000 and
$50,000 and
prompt payment history

discount is 10%.

CASE 4: previous year's purchases between $10,000 and
$50,000 and
delinquent payment history

discount is 5%.

CASE 5: previous year's purchases < $10,000

no discount.

Figure 74: Structured English representation for customer discount policy

Sequence is the simplest of the three constructs to derive; the actions to be performed are listed in order. The framework for the sequence construct is taken from the example that follows.

Figure 73 in Chapter 7 is a data flow diagram of the Inventory Control and Purchasing function at First City Computers. Process 3.4 in Figure 73, PREPARE PURCHASE ORDER, is a function performed when goods are ordered from suppliers. The structured English statements to describe this process are

PREPARE PURCHASE ORDER

Enter supplier name and supplier address on purchase order form.
Enter supplier product code into item field of purchase order form.
Enter reorder quantity into quantity field of purchase order form.
Obtain signature of purchasing agent.

Since the actions are performed once, in consecutive order, the general form for the sequence construct is

action 1
action 2
.
.
.

8.2 Decision Tables

When actions are dependent upon combinations of conditions, structured English is not always suitable, for it may become too complex to read easily. Because readability is a major consideration of the functional specification, decision tables and decision trees are used instead in those situations.

8.2.1 Definition

A *decision table* is a tabular representation of the factors to be considered (the conditions) in making a decision, the actions to be taken when a certain combination of conditions exists, and the rules relating the specific combinations of conditions and actions.

The structured English example using the case construct from the previous section can be expressed as a decision table as shown in Figure 75.

	R_1	R_2	R_3	R_4	R_5
Previous year's purchases (p)	≥ \$50,000	≥ \$50,000	\$50,000 > p p ≥ \$10,000	\$50,000 > p p ≥ \$10,000	< \$10,000
Payment history	prompt	delinquent	prompt	delinquent	N/A
Discount	20%	15%	10%	5%	0%

Figure 75: Decision table for customer discount policy

8.2.2 Types of Decision Tables

There are several different kinds of decision tables: extended-entry, limited-entry and mixed-entry tables; and all have in common the components of factors, actions, conditions and rules. They differ in the content and format of conditions and actions. Examples of each type of table follow.

The example shown in Figure 75 is called an extended-entry decision table. The upper and lower boxes at the left, called the factor and action stubs, respectively, contain a list of the factors and actions. The upper and lower boxes at the right, called the condition and action entries, respectively, contain a list of specific conditions and actions. The columns labelled R_1 through R_5 contain the rules that relate specific combinations of conditions to certain actions. In an extended-entry decision table, each condition entry denotes a specific condition relating to a factor. Each action entry tells the reader what action to perform.

Consider the following example, written in structured English, of the ski school policy at Vail, Colorado:

IF the student has had some experience skiing,
    THEN the student is not required to attend
        the film regarding safety on the slopes.

IF the student has never skied but has skated,
THEN the student is not required to attend the film regarding safety on the slopes.
OTHERWISE, the student must view the film before the first ski class.

The limited-entry decision table for this narrative is shown in Figure 76.

	R_1	R_2	R_3
Experience skiing	Y	N	N
Experience skating	-	Y	N
Attend film	-	-	X
Attend class	X	X	X

Figure 76: Ski school policy

In limited-entry decision tables, the condition entries indicate whether or not a factor exists. The action entries indicate whether or not an action should be taken. The conditions in a limited-entry table are limited to **Y** or **N**, meaning the factor (condition) does/does not exist, or -, meaning the existence of the factor is irrelevant. In this example, if the student has had experience skiing, then any skating experience is irrelevant. The actions entries in a limited-entry table are indicators of whether to perform (**X**) or not to perform (-) the required action.

Figure 77 shows a mixed-entry table equivalent to the limited-entry table shown in Figure 76. Mixed-entry tables have features of both limited-entry and extended-entry tables. As with an extended-entry table, the action entries tell the reader what action to perform. The condition entries, however, are the same as for limited-entry decision tables: **Y**,**N**, and -.

	R_1	R_2
Experience skiing or skating	Y	N
Action	attend class	attend film and class

Figure 77: Another form for ski school policy

8.2.3 Developing Decision Tables

Although all types of decision tables are useful tools, we will limit the discussion and examples that follow to limited-entry tables. Because of the binary nature of the condition and action entries, limited-entry tables translate readily into programming code. In fact, there are table translators available commercially – programs that generate code from decision tables. Other advantages of decision tables include their conciseness and readability.

Lack of contradiction, completeness and redundancy can be achieved in decision tables by applying procedures developed by Thomas Gildersleeve [5]. Contradiction occurs when the same combinations of conditions result in different actions. In such cases the analyst has misinterpreted the information or an error been made in deriving the table. Completeness means all possibilities have been accounted for. Redundancy occurs when a rule is listed more than once.

There are several steps in building the best decision table from a narrative. For our purposes *best* means a decision table that is complete, has no redundancies or contradictions, and has the minimum number of rules possible. We term it a *minimal* limited-entry table.

Gildersleeve's text on system analysis [6] provides the basis for the discussion of decision tables that follows.

The steps involved in building a decision table from a narrative are

1. List the conditions and actions using standard language.
2. Eliminate duplicates.
3. Put conditions and actions in table form, relating the actions to the appropriate combination of conditions.
4. Eliminate redundancy.
5. Rearrange the rules to minimize searching through the table.
6. Check for completeness.

These steps will be discussed using our First City Computers case study.

Recall the description of the Accounts Receivable function from Chapter 6. Assume that after interviewing several employees of First City, the analyst derived the following description.

Prepare Customer Invoice

Invoices are prepared from one copy of the billing slip received from order processing. The customer account file contains information relating to the account, such as billing address, whether or not the customer is exempt from Federal Sales Tax (FST exemption code), and whether or not the customer has preferred status and is therefore entitled to a 10 percent discount.

The amount of the order must be $25 or more, otherwise the terms are COD, except for preferred customers whose account is less than 90 days old. Their terms are 2 percent discount if paid within 10 days and net 30 otherwise (2 percent 10, net 30). Terms are generally 2 percent 10, net 30 unless the customer's account is over 90 days old, in which case the terms are net 30 for preferred customers and COD to regular customers.

In addition to discounts based on the entire shipment, there may be discounts for individual items. The price of each item is obtained from the company price list, and each item has an associated minimum discount quantity. Customers ordering at least the discount quantity, obtain a 5 percent discount off the list price of the item unless they are preferred customers, in which case they get 10 percent. Federal sales tax of 12 percent is added to the non-exempt customers.

Notice that we have to consider the customer, the overall order and the individual items on the order. We will deal with the individual items on the order first. The conditions are

1. quantity shipped greater than or equal to discount quantity (Y or N)

2. preferred customer (Y or N)

and the actions are to give

1. no discount

2. discount = 5 percent

3. discount = 10 percent

Figure 78 shows a table corresponding to the above conditions and actions. We perform this procedure for each item on the shipment, and when finished, we calculate the total and add tax, if applicable. The steps in this process are documented in Figures 79, 80 and 81. We repeat performing the decision table (REPEAT) until there are no more items left; then we EXIT.

	R_1	R_2	R_3
quantity shipped >= discount quantity	Y	Y	N
preferred customer	N	Y	-
no discount			X
discount = 5%	X		
discount = 10%		X	

Figure 78: Decision table for customer discounts

	R_1	R_2	R_3	R_4
more items	Y	Y	Y	N
quantity shipped >= discount quantity	Y	Y	N	-
preferred customer	N	Y	-	-
no discount			X	
5% discount	X			
10% discount		X		
perform TOTAL CALCULATION				X
perform FST CALCULATION				X
EXIT				X
REPEAT	X	X	X	

Figure 79: CALCULATE INVOICE ITEMS

	Choose the applicable case:
calculate shipment total	X
EXIT	X

Figure 80: TOTAL CALCULATION

	R_1	R_2
FST- exempt	Y	N
add FST		X
EXIT	X	X

Figure 81: FST CALCULATION

TOTAL CALCULATION and FST CALCULATION are two subtables within CALCULATE INVOICE ITEMS. Notice that TOTAL CALCULATION is an action table only; it contains no conditions or rules.

We now deal with calculating the terms of the invoice. The conditions are

1. account > 90 days

2. preferred customer

3. shipment total >= $25

and the actions are

1. terms = COD

2. terms = 2 percent 10, net 30

3. terms = net 30

The corresponding table is shown in Figure 82.

	R_1	R_2	R_3	R_4	R_5	R_6	R_7
account > 90 days	N	Y	Y	N	N	Y	Y
preferred customer	Y	Y	N	N	-	Y	N
shipment total > minimum	N	N	N	N	Y	Y	Y
terms = COD		X	X	X			X
term = 2% 10, net 30	X				X		
terms = net 30						X	

Figure 82: DETERMINE TERMS

Rules 2 and 3 are *dependent*, meaning they differ by one condition entry and they have the same set of actions. As dependent rules, they may be combined and replaced by rule 2′. R_4 through R_7 will now be called R_3 through R_7 as shown in Figure 83.

	R_1	R_2'	R_3	R_4	R_5	R_6
account > 90 days	N	Y	N	N	Y	Y
preferred customer	Y	-	N	-	Y	N
shipment total > minimum	N	N	N	Y	Y	Y
terms = COD		X	X			X
term = 2% 10, net 30	X			X		
terms = net 30					X	

Figure 83: DETERMINE TERMS - refined

The table DETERMINE TERMS is a minimal limited-entry decision table. The table is complete because all possible combinations of conditions have been accounted for. The number of possible rules in a limited-entry table is 2^N, where N = the number of conditions. The table DETERMINE TERMS has three conditions, so the number of possible rules is $2^3 = 8$. (Rules R_2' and R_4 are combinations of two rules each.)

For each invoice, the individual items are calculated, and the terms are determined. This may be shown as the decision table of Figure 84.

perform CALCULATE INVOICE ITEMS	X
perform DETERMINE TERMS	X

Figure 84: PREPARE CUSTOMER INVOICE

It is easy to see that decision tables are a precise, readable tool for both the user and systems personnel. Together Figures 79, 80, 81, 83 and 84 describe the process PREPARE CUSTOMER INVOICE. The data dictionary entry for PREPARE CUSTOMER INVOICE is shown in Figure 85.

PROCESS NAME: PREPARE CUSTOMER INVOICE

PROCESS NUMBER: 2.3

PROCESS DESCRIPTION:

PREPARE CUSTOMER INVOICE

perform CALCULATE INVOICE ITEMS	X
perform DETERMINE TERMS	X

CALCULATE INVOICE ITEMS

	R_1	R_2	R_3	R_4
more items	Y	Y	Y	N
quantity shipped >= discount quantity	Y	Y	N	—
preferred customer	N	Y	—	—
no discount			X	
5% discount	X			
10% discount		X		
perform TOTAL CALCULATION				X
perform FST CALCULATION				X
EXIT				X
REPEAT	X	X	X	

DETERMINE TERMS

	R_1	R_2'	R_3	R_4	R_5	R_6
account > 90 days	N	Y	N	N	Y	Y
preferred customer	Y	—	N	—	Y	N
shipment total > minimum	N	N	N	Y	Y	Y
terms = COD		X	X			X
terms = 2% 10, net 30	X			X		
terms = net 30					X	

TOTAL CALCULATION

calculate shipment total	X
EXIT	X

FST CALCULATION

	R_1	R_2
FST- exempt	Y	N
add FST		X
EXIT	X	X

Figure 85: Data dictionary entry for process 2.3

8.2.4 Minimal Decision Tables

The procedures for eliminating redundancy within decision tables warrant more explanation. The goal is to produce a *minimal* table, a table with the minimum number of rules. To do this, all redundancies between rules must be eliminated so that the rules are independent of each other. A rule is called *simple* if all the condition entries are either **Y** or **N**; if any condition entries are -, the rule is called *compound*. Consider the decision table shown in Figure 86 in which all the rules are simple. In order to minimize the table, we examine pairs of rules to find any that contain the same action entries but one different condition entry (i.e., dependent pairs). R_1 and R_4 qualify as a dependent pair of rules. They can be combined, as shown in Figure 87.

	R_1	R_2	R_3	R_4	R_5	R_6	R_7	R_8
C1	Y	Y	N	Y	Y	N	N	N
C2	Y	N	N	Y	N	Y	Y	N
C3	Y	N	N	N	Y	N	Y	Y
A1		X			X	X		
A2	X		X	X				X
A3							X	

Figure 86: Decision table

C1	Y
C2	Y
C3	-
A1	
A2	X
A3	

Figure 87: Rule reduction

Further reductions lead to the minimal, sorted table in Figure 88. The rules in the table shown in Figure 88 have been rearranged (and renamed) to minimize the effort required when searching the table for a particular rule. The sorted table was derived by putting all the rules for which condition 1 (C1) is Y to the left and all the rules for which condition 1 is N to the right. Then condition 2 (C2) was examined, and the rules for which both C1 and C2 are Y are placed to the far left, followed by the rules for which C1 is Y and C2 is N. Next come the rules for which C1 is N and C2 is Y. Finally the rules for which both C1 and C2 are N are placed to the right. This procedure is repeated with condition C3 taken into consideration. (This is analogous to sorting in computer programming: C1 would be the major sort key, C2 would be the second sort key and C3 would be the minor sort key.) The resulting left-to-right order of the rules is

1. Rule for which C1, C2 and C3 are Y
2. Rule for which C1 and C2 are Y, and C3 is N
3. Rule for which C1 is Y, C2 is N, C3 is Y
4. Rule for which C1 is Y, C2 and C3 are N
5. Rule for which C1 is N, and C2 and C3 are Y
6. Rule for which C1 is N, C2 is Y, C3 is N
7. Rule for which C1 and C2 are N, C3 is Y
8. Rule for which C1, C2 and C3 are N.

When sorting decision tables, conditions having irrelevant entries are considered last.

	R_1	R_2	R_3	R_4	R_5
C1	Y	Y	N	N	N
C2	Y	N	Y	Y	N
C3	-	-	Y	N	-
A1		X		X	
A2	X				X
A3			X		

Figure 88: Minimal, limited-entry decision table

8.3 Decision Trees

Decision trees are an alternative tool for describing process logic. They are especially appropriate when the numbers of conditions and outcomes are relatively small. However, their usefulness goes beyond software development and into management science.

8.3.1 Definition

A decision tree is a graphic of the decisions, events or chances, and the outcomes associated with the decisions and chances. The decision table of Figure 78 is shown in decision tree form in Figure 89. Nodes E_1 and E_2 are event or chance nodes, and O_1, O_2 and O_3 are the outcomes. Notice that there are no decisions involved in the example shown in Figure 89, only chances or events.

The next example, which is more general in nature, illustrates all three components of a decision tree.

Suppose you have been approached by colleagues who wish to start their own consulting business and would like you to join. Each person will contribute $25,000 toward office space and equipment, computer hardware and software, salaries and so on. If successful, the business could generate an annual income of $80,000 for each of the principals. If unsuccessful, all could be lost. Your current annual salary is $35,000.

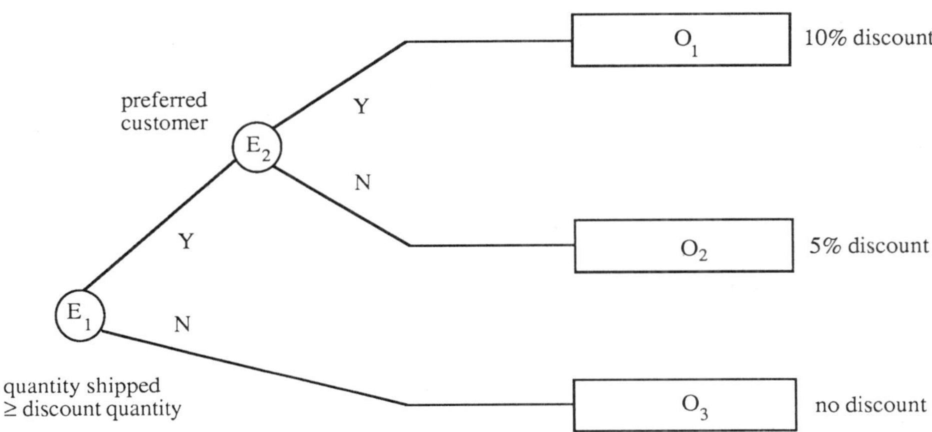

Figure 89: Decision tree form for customer discount policy

The decision tree for this situation is shown in Figure 90.

D_1 is the conscious decision you have to make – whether or not to join the consulting firm. If you reject the offer to join, you gain $35,000 by staying at your current job. However, if you decide to join the firm, one of two events may occur: the business may be successful and you can expect to earn $80,000, or it may fail and you stand to lose your $25,000. (In fact, there are many possibilities in between that have not been considered.)

Decision trees are effective when the number of decisions and events is relatively small, although any logic that can be represented by a decision table can also be represented by a decision tree.

8.3.2 Developing Decision Trees

During the programming phase of software development, code can easily be written from a decision tree, and by judiciously sequencing the events in a decision tree, processing efficiency may be improved without affecting the clarity of the logic involved.

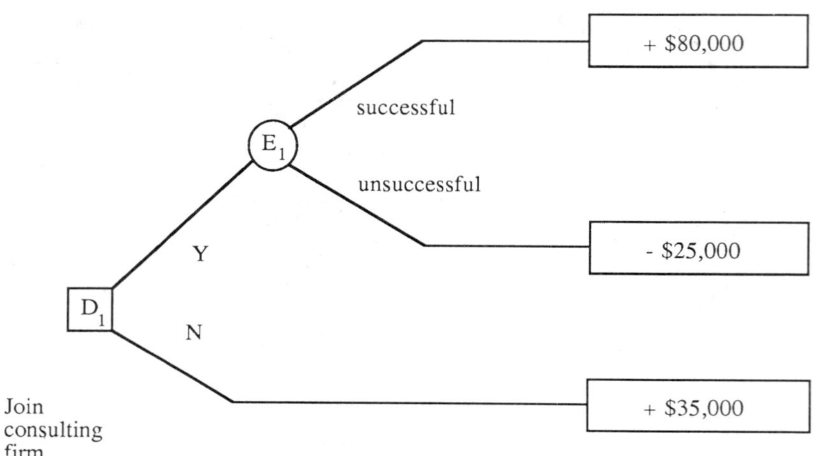

Figure 90: Decision tree for consulting firm decision

For example, suppose a national airline has two classes of commercial customers, preferred (and as such entitled to discounts) and non-preferred. Preferred customers are those who have booked 5,000 or more seats during the past year and who also fulfill either one of the following conditions: their account is in good standing, or the most current Dun and Bradstreet report on them is favorable. The decision tree for this example is shown in Figure 91.

Now suppose our commercial customer file has 100,000 records, and through sampling we have determined that

20 percent of the accounts book 5,000 or more seats annually,

90 percent of the accounts have acceptable D&B reports,

70 percent of the accounts are in good standing.

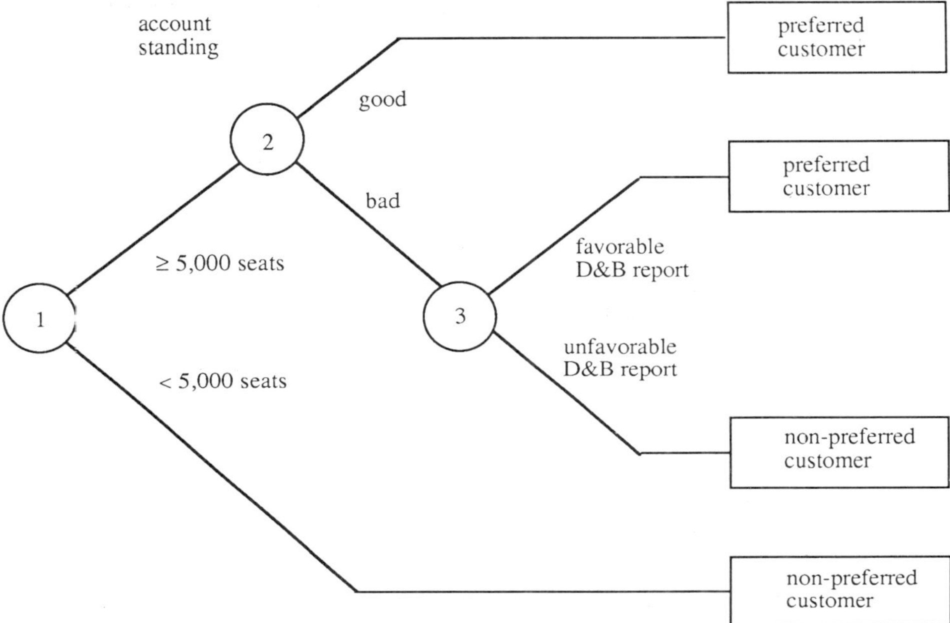

Figure 91: Airline procedure for determining preferred customers

To determine how many tests of the file are necessary in order to obtain a list of preferred customers, we proceed as follows:

The first test, #1, tests all 100,000 records.

The second test, #2, tests 20 percent of 100,000 or 20,000 records.

The third test, #3, tests 70 percent of 20,000 or 14,000 records.

So 134,000 tests on the file are necessary in order to obtain a list of preferred customers using the above sequence of tests.

Consider the alternative sequence shown in Figure 92 for the same procedure. The number of tests required for this arrangement is 227,000, considerably more than the first arrangement of tests. Computer processing time can be improved by careful sequencing of tests, while preserving the clarity of the logic.

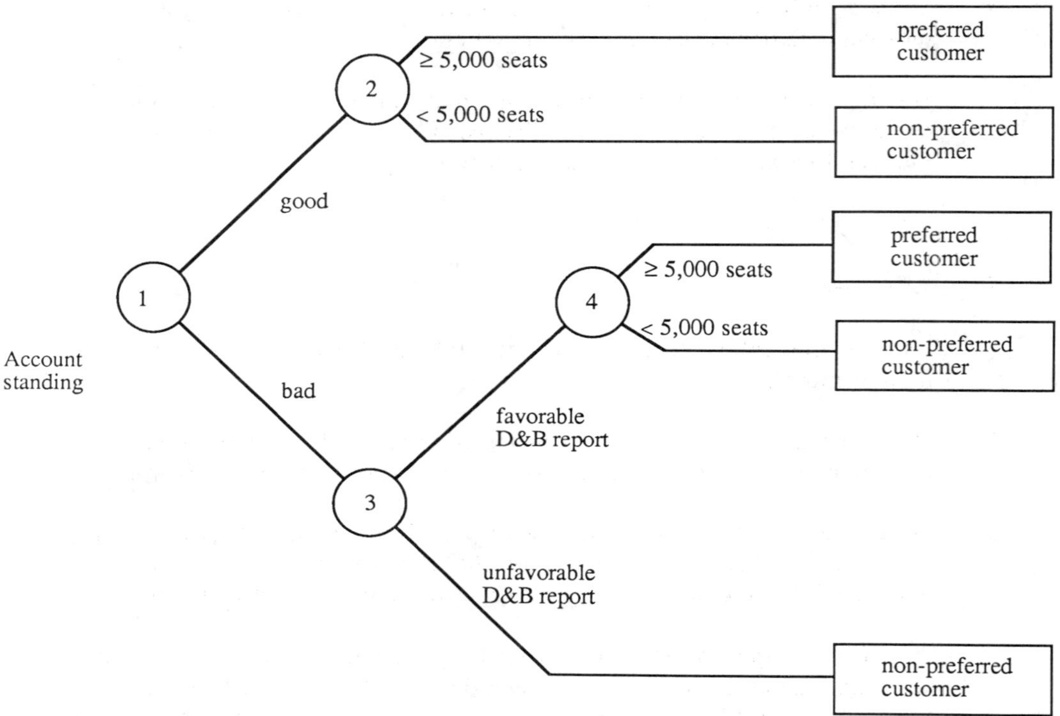

Figure 92: Alternative sequence

8.4 Process Logic - Summary

The bottom-level processes in the DFD's of the system conceived during analysis to satify users' business needs become program modules. One of the tasks of the analyst is to specify the logic of each of these processes. Three commonly-used tools for modeling process logic are structured English, decision tables and decision trees. All three tools are nontechnical enough to be understood by users, and precise enough for programmers to use in translating the specification to computer programming code. Each tool is suitable for different situations:

- Structured English is suitable for expressing relatively uncomplicated logic.

- Decision tables are best used for expressing processes that involve complex combinations of conditions and actions.

- Decision trees are best for expressing combinations of conditions and outcomes when there are relatively few conditions and outcomes (otherwise the tree becomes too complicated to read easily).

8.5 Exercises

1. Express the extended-entry decision table of Figure 77 as a minimal limited-entry table.

2. Develop a guideline for the sequencing of tests for decision trees when the probabilities of the outcomes to each test are known or can be estimated.

3. Develop a (set of) minimal limited-entry decision table(s) for the following narrative in which the secretary of the Department of Ersatz Sciences describes how students are accepted into the Master's Degree program.

   "We get applications from all kinds of students from all kinds of universities. Therefore, we have ranked different schools as A, B or C, where A are those schools that graduate the best students. If students from an A school apply, they are almost certainly admitted to the program, unless of course, their averages are below C-, in which case we wait and see how good their letters of recommendation are. Graduates from B schools with averages above B+ are admitted without question. Decisions for those from a B school with an average of B+ or below, or from a C school, are based on the letters of recommendation. The letters are examined by three faculty members who use a numerical code and assign a number (1, 2 or 3) to the set of recommendation letters of each applicant. A good set of letters is given a 3, and a bad one a 1. If the sum of the code numbers assigned by the faculty for a given applicant is 7 or higher, the student is admitted. The exceptions to this system are the applications of students from unranked universities; for these students to be admitted, their average has to be above B+ and the sum of the code numbers 7 or higher. Of course, students are not admitted unless there are faculty members willing to act as their thesis supervisors, and in addition, all foreign students must have a passing grade in the English Language Proficiency Exam (ELPE) before being admitted. As you can see, the procedure is very simple."

4. Express the logic of Exercise 3 in structured English form.

5. Express the logic of Exercise 3 in decision tree form.

6. A computer program uses three switches, S_1, S_2 and S_3, each of which show the values 0 or 1. An action must be taken according to the value of W, where $W = S_1 - S_2 + S_3$. Different actions are required for different values of W.

   a. Construct a limited-entry decision table for this problem with conditions on the value of W.

   b. Find the probability of each rule, and construct the best decision tree by choosing first the rules with the highest probabilities.

7. A large insurance company has an employee pension plan. The plan has three levels of service: Tiers 1, 2 and 3. Tier 1 provides the basic pension benefit at no cost to the employee. As soon as a full-time or part-time employee has reached three months of employment, plan membership begins automatically.

   The company's contributions to Tier 1 are "fully vested" after five years of service. This means that all pension credits are retained in the name of the employee, even upon the employee's leaving the company. If the employee leaves before two years of service have been completed, no credits are vested. Vesting begins at the rate of 10 percent after two years service and increases by 2.5 percent for each succeeding month until five years of service have been reached.

   Tier 2 provides the employee with an opportunity to enhance the Tier 1 pension by making contributions via payroll deductions. These contributions are fully matched by the company. The employee's and the company's contributions are vested immediately.

   Tier 3 allows the employee to further supplement the Tier 1 and Tier 2 pension by additional contributions, with no company matching.

   In Tier 1, if the employee is retiring, the following actions are taken:

- For service prior to January 1, 1984:

Pension benefits are calculated according to the terms of the plan in force at that time.
If an employee choses to retire prior to age sixty-five, this portion of the pension is reduced by 0.5 percent for each month that early retirement occurs in advance of age sixty-three.

- For service after December 31, 1983:
  Assuming that a person retires at the normal retirement age, the pension for post-1983 service is calculated on the basis of the final three years average earnings and the years of service.
  If a person chooses to retire prior to age sixty-five, this portion of the pension is reduced by 0.5 percent for each month that early retirement occurs in advance of age sixty-five.

In both cases, the accumulated funds are used to purchase a form of lifetime retirement income.

If an employee leaves the company prior to retirement, the value of the Tier 1 pension credits (subject to vesting) will be transferred to Tier 2 for investment until retirement.

All Tier 2 contributions (if any) are invested until the employee retires and then the accumulated funds are used to purchase a form of lifetime retirement income.

Tier 1 and 2 funds cannot be removed from the company before retirement. However, these funds can be transferred to another employer.

According to current federal law, Tier 3 contributions cannot be withdrawn in cash prior to retirement, unless a person is leaving a current employer and will not be employed subsequently.

a. Express the above narrative in decision table form.

b. Express the same logic in decision tree form.

c. Express the above narrative using structured English.

d. Which representation is the easiest to understand? Why?

8. The narrative that follows describes the procedure carried out by the clerical staff within the warehouse of an electrical parts distributor. It was derived after the analyst conducted several interviews with appropriate users.

   When an order is received from the sales department, each item on the order is checked to see if it can be met from our current inventory. If sufficient inventory is held, the warehouse clerk adjusts the stock records and passes the item for picking and dispatch. Each time the stock records are changed, the new inventory level is compared to the reorder level that is marked on the stock-item card. If the new inventory level is below the reorder level, the warehouse clerk writes out a purchase order form, notes the quantity ordered on the stock-item card, and passes the purchase order form to the chief buyer for approval and dispatch to the supplier.

   If there is some inventory, but not enough to fill the order, the warehouse clerk dispatches the items available, adjusts the stock records, and creates a back order for the required amount. The back order is filed in part-number order, awaiting receipt of a shipment. If the item has been previously ordered, the warehouse clerk sends an expedite-delivery notice to the chief buyer. If the item is not on order, a purchase order form is prepared.

   If the item is completely out of stock, a back order is created and filed as above, and the clerk sends an expedite-delivery notice to the chief buyer whether or not the item is already on order.

   a. Express the above logic in structured English.

   b. Express the narrative in decision table form.

   c. Express the above logic in decision tree form.

9. Suppose that the number of courses required by your university or college to complete a degree in computer science has been changed. Assume that the requirement until now has been that a student must have completed forty-eight semester courses (all courses carry the same weight). Starting next term, students need to complete forty-four semester courses to obtain a degree in computer science.

The student registration system is an automated system that schedules student classes, and processes their grades, academic standing and fees. It must be changed to accommodate the revision to degree requirements.

Computer science students have been required to maintain overall and math averages of 65 percent; otherwise, they have been placed on probation. Students are required to withdraw from a computer science major program if they have had two probationary terms. The terms of probation, failure, and required withdrawal from computer science under the new degree requirements will be changed as follows:

A passing term is defined as an average of greater than or equal to 65 percent on all required subjects, provided the student has not failed a subject. If the student has failed a required course, the student's average must be an additional 5 percent higher for each course failed. A student may not fail more than two required courses in any one term. A student failing an elective course must have an average (over all required courses) that is 3 percent higher than before. A student may not fail more than three courses in any one term. A student receiving less than a 65 percent average on the required courses, but more than 60 percent and having no failures, will not pass, but will be placed on probation. If a student has failed one elective and has an average of between 60 percent and 68 percent, the student will be placed on probation. A student failing one required course and having between 60 percent and 70 percent, will be eligible for probation. No student who has failed more than one course (either required or elective) and who doesn't meet the passing standard will be placed on probation. A student will be required to withdraw from Computer Science for receiving a failed term, or two consecutive probationary terms.

Express the above logic in decision table form.

8.6 Bibliography

1. Brown, David B. and Herbanek, Jeffrey A. *Systems Analysis for Applications Software Design.* Oakland, California. Holden-Day, Inc., 1984.

2. Curtis, Bill (editor). *Tutorial: Human Factors in Software Development.* Los Angeles, California. IEEE Computer Society, 1981.

3. DeMarco, Tom. *Structured Analysis and System Specification.* Englewood Cliffs, New Jersey. Prentice-Hall, 1979.

4. Gane, Chris and Sarson, Trish. *Structured Systems Analysis: Tools and Techniques.* Englewood Cliffs, New Jersey. Prentice-Hall, 1979.

5. Gildersleeve, Thomas R. *Decision Tables and Their Practical Application in Data Processing.* Englewood Cliffs, New Jersey. Prentice-Hall, 1970.

6. Gildersleeve, Thomas R. *Successful Data Processing System Analysis, Second Edition.* Englewood Cliffs, New Jersey. Prentice-Hall, 1985.

7. Parkin, A. *Systems Analysis.* London, U.K. Edward Arnold, 1980.

8. Pressman, Roger S. *Software Engineering: A Practitioner's Approach.* New York, New York. McGraw-Hill, 1982.

9. Senn, James A. *Analysis and Design of Information Systems.* New York, New York. McGraw-Hill, 1984.

10. Weinberg, Gerald M. *Rethinking Systems Analysis and Design.* Boston, Mass. Little, Brown and Company, 1982.

Chapter 9

CONSOLIDATION INTO A BUSINESS CASE

Chapter 6 describes in detail the six major phases of analysis. The fourth phase requires that the analyst develop several strategies for automation and present a business case for each (costs, benefits, timing, priority, etc.). In this chapter we explore the costs and benefits pertinent to software projects in order to assess the economic feasibility of each strategy. Our main concern in this discussion is economic feasibility; we will not be considering other issues such as technical feasibility or operational feasibility, or more nebulous considerations that defy quantification. In addition to a cost/benefit analysis for each automation strategy, a more detailed analysis of the selected option is often required in order to obtain corporate approval for further development and implementation.

9.1 Why Discuss Costs and Benefits?

There is a difference between spending money and investing money. We spend money on clothing, food, shelter and other goods and services necessary for our existence. For the most part, when we spend money on these kinds of things we expect no monetary benefit. However, when investing money, we expect to receive something in return. If we invest in savings bonds or term deposits we expect to receive interest. If we buy stocks we anticipate dividends or an increase in value. When buying real estate we hope its worth will appreciate.

To a company, a new system represents an investment. Funds will be committed throughout the development and life of the system, and in return, the company, or more properly, its management, expects to receive benefits. These benefits may take the form of faster processing of customer orders and outgoing invoices, and faster collection of accounts, with a consequent savings on bank interest on the company loan. The benefit could take the form of providing more and better information, allowing management to do a better job. An example is better management control of inventory through the system's providing management with annual turnover rates and corresponding relative profitability for each item. In cases such as this, the benefits of automation may be hard to estimate and are often referred to as intangible benefits. Regardless of the form the benefits may take, they are always anticipated in a software development project.

Companies develop systems as investments, and if costs exceed either the real or perceived value of the benefits, then the system is not economically feasible, and represents a bad investment. It is the responsibility of the analyst to provide management with estimates of all costs and benefits pertaining to each possible option so that they can assess its economic feasibility.

The salaries of the analysts, designers, programmers and others working on the project during development represent a cost to the company. After the system has been installed, there will be other costs pertaining to its operation and maintenance. All of these costs must be considered when assessing the cost of software development.

Typically, companies have many more opportunities for investing than they have capital to finance the opportunities. How do they pick the right investments? When supplied with all of the information possible with respect to an investment opportunity, management can calculate which opportunity is the right one. From their point of view, an investment in a new system is no different from any other capital investment, and if the potential benefits do not exceed the costs, then the system should not be developed. The document containing all the information about a proposed investment is called a *business case*. All investments, from software to facilities to people, require management approval of their associated business case.

9.2 Concepts and Terms

Before proceeding, we need to define some commonly used investment terminology.

First of all, a *cost/benefit analysis* is the process of isolating and estimating the costs and benefits pertinent to a proposed software system. Cost/benefit analyses are often required by organizations for software systems development, market research and quality assurance projects. As we shall see, the costs are normally much easier to estimate or quantify than the benefits.

Benefits may be of a qualitative nature, nevertheless, the analyst must attempt to objectively quantify them. For example, one result of a proposed office automation system would be the upgrading of skills of office employees who will use the system. From their perspective, the upgrading may be an important benefit: their skills would be more marketable, and prove valuable to career development. However, is this a benefit for the company, and if so, how can its value to the company be quantified? Management may perceive upgrading the skills of its workers as a cost: employees with increased skills demand higher salaries.

Costs and benefits of investments are related to time. For example, suppose a company is considering purchasing a delivery truck to add to its fleet of vehicles. The purchase price and financing charges are costs, but they are not the only ones borne by the company. The insurance premiums, the vehicle's depreciation, and cost of gasoline, repairs and maintenance are also costs incurred during the life of the truck. These costs which pertain to the day-to-day operation of the truck, are called *operating costs.*

The benefits provided by the truck also relate to the period of time the truck will be in service for the company. Having an additional truck may provide faster delivery to customers than that provided by the competition, and it may result in more business. If the company has several warehouses, the truck may also be useful for transporting inventory between warehouses, thereby saving shipping costs. Both the savings in shipping costs and the increase in business represent benefits derived by adding the truck to the company's fleet.

The period of time during which an investment is expected to be of value (expected to provide benefits) to a company is called its *economic life*. Before the costs and benefits of a proposed system, or of any potential investment, can be determined, the economic life of the investment must be estimated.

For some investments, such as land or buildings, the economic life may be indefinitely long. For others, such as machinery or vehicles, the economic life is relatively short and easily determined. Because analysis is concerned with software systems development, we are concerned with the economic life of the software only. (Implicit in this concern is that the economic life of the hardware is at least as long as the software. Even though we are not concerned with the economic life of the hardware, we are concerned with the hardware costs associated with a proposed software system – hardware that must be purchased, the existing hardware resources required for development and operation of the system – costs that must be isolated and estimated.)

It was indicated in previous chapters that the average life of a software system is five to seven years. For the examples given in this chapter, we will assume an economic life of five years. The conservative end of the average is chosen for several reasons. First of all, to estimate an investment's economic life is to project into the future. With each additional year the accuracy of the projection becomes more questionable: changes in the economic climate, in government regulations, in position in the marketplace or in technology may have drastic effects on anticipated benefits. As well, if later it turns out that the system's economic life has been overestimated, the analyst may be held accountable by management. However, to underestimate anticipated benefits does no harm. For all of these reasons, it is best to conservatively estimate a system's economic life, hence the choice of five years.

Now that the economic life of the potential investment has been determined, we are in a position to define some commonly used terminology pertaining to costs.

Fixed costs are costs that do not vary with the volume of activity (the amount of business done by a company). Fixed costs – property taxes, mortgage payments, rent and insurance, for example – are costs that are automatically incurred with the passage of time, and they remain unchanged by the level of success achieved by the business.

Variable costs are those costs that vary as the volume of business changes. For individuals, income tax is an example of a variable cost because the more money one earns, the more tax one pays. Sales commissions, supplies and even salaries are examples of variable costs for organizations. Salaries represent a variable cost because when the economic climate for a company is unfavorable, costs can be reduced by laying off or terminating employees. Conversely, when business is good, organizations increase costs by purchasing more supplies and services, hiring more employees and paying more in sales commissions. Because ultimately almost all costs are variable, fixed costs are always related to a period of time. For example property taxes may remain unchanged for several years and then be reassessed. Rent usually remains constant during the lease period but thereafter may change.

Variable costs pertaining to a specific situation are called *differential* or *relevant costs*. These are costs that occur only because of a specific investment opportunity; such costs differ among alternatives. Costs pertaining to investing in a new product line are different from the costs pertaining to developing a new software system, both of which may be investment alternatives. *All* costs pertaining to a systems analysis proposal are differential costs. Salaries and company benefits for the personnel involved are differential costs, as are the purchase of software and hardware, and the computer resources required for running the proposed system. These differential costs are estimates of what one expects the development and operation of the system to cost.

We define one last investment term. A *sunk cost* is a cost that has already been incurred. It exists because of past actions and cannot be changed, regardless of the actions we take. An example of a sunk cost would be the cost of an obsolete piece of equipment that cannot be sold. Another example would be the cost of a purchased software package which proved to be inappropriate, and therefore unused. Even if the software package had been purchased to solve the user's current business needs, its cost does **not** pertain to the cost of the system now under consideration – it represents a sunk cost to the company as a whole, but not to the proposed software system. In systems development there are no sunk costs: all costs are differential and must be considered.

9.3 Profile of an Investment

The life of an investment typically has a development phase followed by an operating period. During the development phase, costs that are necessary to create the product are incurred and are called *development costs*. During the operating period, costs required to ensure the product's continued existence are incurred and are called *operating costs*. For a proposed software system possible development costs include salaries and company benefits of the systems analysts, users, programmers, consultants and others who work on the development; computer resources required for testing; training expenses; and any purchased software. For other investments, such as the establishment of a new branch sales office or the development of a new product line, development costs may include the refurbishing of facilities, promotional advertising and the cost of relocating employees. The costs or cash outflows continue throughout the investment's development period.

Eventually the development phase concludes and the operational portion of the investment begins with its associated operating costs and benefits. In the case of a proposed system, this happens when the system has been implemented and is in production. For the corporation opening a branch sales office, the operating phase begins when the office has opened, and for the company developing a new product line, it begins when the products have been produced (or acquired) and are available to customers.

In the operational phase, benefits or revenues begin to accrue, probably slowly at first. There may still be ongoing development, but at some point the development costs will be phased out by the operating costs. For a software system, operating costs include computer resources, supplies such as computer forms and paper, and the salaries and benefits of maintenance programmers, of computer operators and of data entry personnel. If the company has made a good investment, the cumulative benefits eventually exceed cumulative development and operating costs.

Toward the end of an investment's economic life, the benefits may decrease and operating costs rise. Consider again the example of the delivery truck from the previous section. Initially vehicles are relatively service-free, however, with age and use, maintenance costs increase and eventually the vehicle may completely wear out or require repairs so expensive that buying a replacement is more economical.

For computer systems, a similar scenario can occur. The system provides benefits throughout its economic life, but because of changes in the business environment or advances in technology, the system's benefits decline, maintenance costs increase, and eventually a replacement is required. This situation is illustrated by the cash flow table shown in Figure 93.

Suppose that the investment shown in Figure 93 has a two-year development period (years 1 and 2) followed by a five-year economic life (years 3 through 7). Its seven-year life is shown across the table as column headings. The row headings represent the development costs, operating costs, receipts and the cumulative balances of receipts less costs for each year. At a point between years 2 and 3 the cumulative cash outflow begins to diminish. Then, in between years 4 and 5 the balance changes from negative to positive.

	($000)						
	Year 1	Year 2	Year 3	Year 4	Year 5	Year 6	Year 7
Beg. Bal.	0	(20)	(55)	(54)	(20)	33	98
Devel. Costs	20	35	10	0	0	0	0
Operating Costs	0	0	12	13	15	17	18
Receipts	0	0	23	47	68	82	76
Ending Bal.	(20)	(55)	(54)	(20)	33	98	156

Figure 93: Cash flow table

9.3.1 Methods of Analyzing Investments

Just as a system may be viewed or analyzed from different perspectives such as function, operating mode, and communications environment (refer to Chapter 2), investments also can be analyzed in different ways. In this section we explore several common methods by which investments can be analyzed.

The information in Figure 93 can be represented as a graph, shown in Figure 94. (This is actually the mathematical integral of the cash flow.) This figure is sometimes called a *profit curve*. The points Y_1 to Y_7 represent years 1 through 7. Point A is the point where the outflow of cash is the greatest. Point B, the point at which net benefits begin to exceed net costs, is called the *break-even point*. Total costs and total receipts are equal at the break-even point.

The period of time between the origin and point B is called the *payback period*. It is the period of time over which the investment's cost will be repaid from the anticipated benefits (provided the estimates are correct). For example, suppose one is considering purchasing a piece of equipment which costs $6,000. The annual benefit (receipts less operating costs) is expected to be $3,000. Then the payback period is two years ($6,000 divided by $3,000). The general formula for the payback period is

$$\text{Payback Period} = \frac{\text{Investment}}{\text{Annual Benefit}}$$

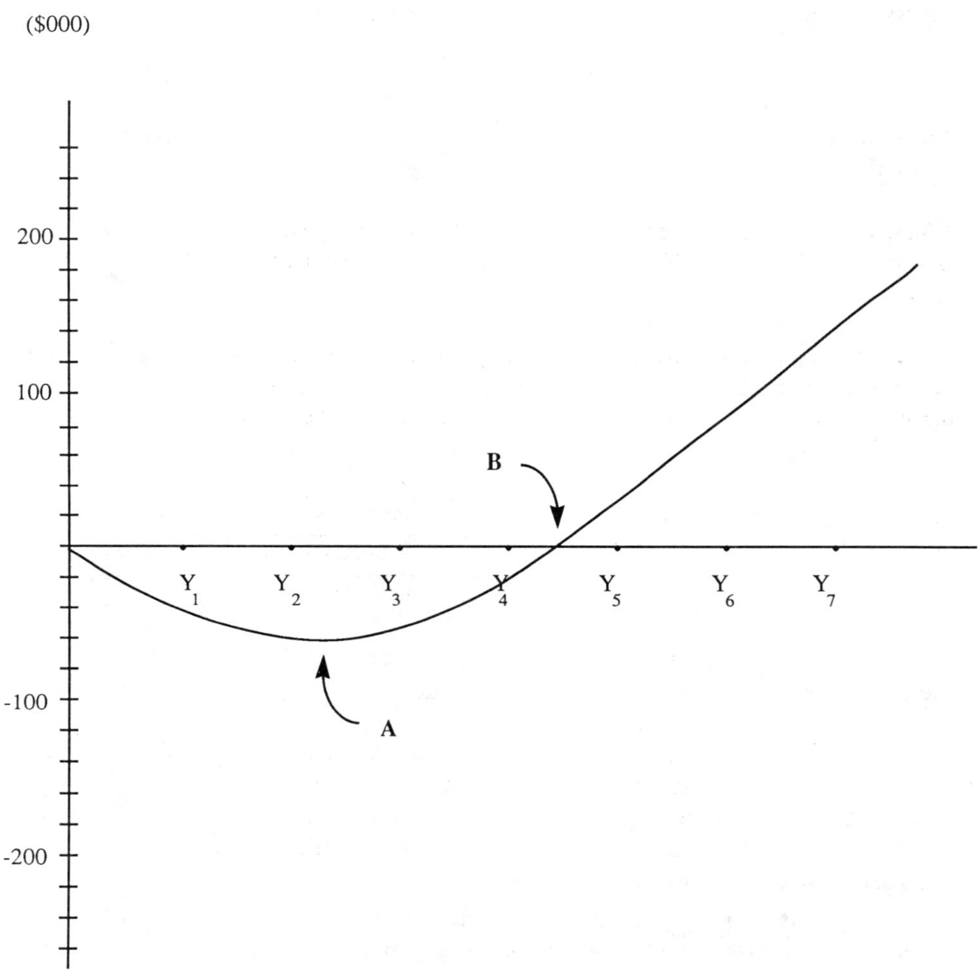

Figure 94: Profit curve

Consider the following example. Suppose we have two potential investments, A and B. The relevant information for each is shown below.

	Project A	Project B
Investment	\$5,000	\$10,000
Annual benefit	\$2,500	\$2,500

Using our formula, Project A has a payback of two years and Project B has a payback of four years. Clearly, within the context of payback, A is the better investment.

Payback is valid only for comparing potential investments having the same general characteristics, and for such cases it is often used as a "quick and dirty" method of analysis. *Benefits* are defined to be receipts less operating costs. Then for a proposed system, the formula for the payback period is

$$\text{Payback Period} = \frac{\Sigma \text{ Development Costs}}{\dfrac{\Sigma\,(\text{Annual Receipts - Annual Operating Costs})}{\text{Economic Life}}}$$

where Σ, the Greek letter sigma, means "the sum of." Note that Σ Development Costs = Investment.

What this formula means in words, is this: the payback period is the sum of all development costs (the investment) divided by the average annual benefit (the sum of the annual receipts less the sum of the annual operating costs divided by the economic life of the system). As an illustration, consider again the cash flow table shown in Figure 93.

$$\Sigma \text{ Development Costs} = \$65{,}000$$
$$\text{Economic Life} = 5 \text{ years}$$
$$(\text{Annual Receipts - Annual Operating Costs}) = \$221{,}000.$$

The average annual benefit = \$44,200, and the payback period = 65,000/44,200 = 1.47 years. (The discrepancy between the value arrived at by the formula and the graph in Figure 93 results from the annual benefit's not being a constant amount. The formula for payback period is used for predicting the future. The cash flow table shown in Figure 93 represents actual data.)

9.3.2 The Concept of Present Value

When considering investments, one anticipates that the investment will be worth more in the future. For example, if a dollar is deposited today into a savings account with an interest rate of 10 percent, the dollar will be worth $1.10 one year from now. Conversely, given an interest rate of 10 percent, $1.10 received one year from now is worth a dollar today. If $.91 is deposited today in a savings account with an interest rate of 10 percent, the $.91 will be worth $1.00 one year from now, and conversely, given an interest rate of 10 percent, $1.00 received one year from now is worth $.91 today. The logic underlying this example can be extracted: a dollar, or any given sum of money is worth more today than the same amount a year from today. It follows, then, that an amount received a year from now is not worth the same amount today, but is worth somewhat less. The value of money today is called its *present value.*

Assume again that one will invest a dollar today at 10 percent interest. At the end of one year, the dollar will be worth $1.10. If the $1.10 remains on deposit for another year, at the end of the second year the investment will be worth

$$\$1.10 \times 1.1 = \$1.21$$

At the end of three years its value will be

$$\$1.21 \times 1.1 = \$1.331$$

9.3.2.1 Compound Interest Formula

The compound interest formula is used to calculate the *future value* of investments. In general, if one is investing P dollars today, and if the annual interest rate is i compounded annually, and the number of years (or compounding periods) is n, then the future value F of an investment is given by the formula

$$F = P(1 + i)^n \qquad \text{(Equation 1)}$$

From Equation 1, the value of an investment of $1.00 (P = 1) at the rate of 10 percent (i = .10) at the end of four years (n = 4) is calculated as follows:

$$F = 1(1.1)^4 = (1.1)^4 = 1.4641$$

The future value of a dollar (P = 1) n years from now at the annual interest rate of i is given by

$$F = (1 + i)^n$$

One anticipates that a proposed system will provide benefits in the future. How are these future benefits converted to their present value?

To do this, the compound interest formula shown in Equation 1 is used. We solve for the variable P.

$$P = \frac{F}{(1 + i)^{n}}$$

To state this result in words, P is the present value of a future benefit F, to be received n years from now at an interest rate of i. By substituting for the future value F, it is possible to find out what the value of an investment, worth one dollar in the future, is worth at present if it is invested for one year at an interest rate of i.

Solving the equation for F = 1, i = 10 and n = 1 year,

$$P = \frac{1}{1.1} = .909$$

we find that an investment worth a dollar one year from today is worth \$.909 presently if invested at 10 percent.

To verify the formula, we can use figures from a previous example of a dollar invested for one year at 10 percent and yielding a future value of \$1.10. Again we are solving for the present value.

$$P = \frac{1.10}{1.1} = 1$$

When converting future dollars to their present value, the interest rate i is called the *discount* rate and the process of converting is called *discounting*.

Figure 95 is a present value table which shows the present value of an investment worth one dollar in future years calculated for various discount rates and investment periods.

Years	1%	2%	4%	6%	8%	10%	12%	14%	15%	20%
1	.990	.980	.962	.943	.926	.909	.893	.877	.870	.833
2	.980	.961	.925	.890	.857	.826	.797	.769	.756	.694
3	.971	.942	.889	.840	.794	.751	.712	.675	.658	.579
4	.961	.924	.855	.792	.735	.683	.636	.592	.572	.482
5	.951	.906	.822	.747	.681	.621	.567	.519	.497	.402
6	.942	.888	.790	.705	.630	.564	.507	.456	.432	.335
7	.933	.871	.760	.665	.583	.513	.452	.400	.376	.279
8	.923	.853	.731	.627	.540	.467	.404	.351	.327	.233
9	.914	.837	.703	.592	.500	.424	.361	.308	.284	.194
10	.905	.820	.676	.558	.463	.386	.322	.270	.247	.162
11	.896	.804	.650	.527	.429	.350	.287	.237	.215	.135
12	.887	.788	.625	.497	.397	.319	.257	.208	.187	.112
13	.879	.773	.601	.469	.368	.290	.229	.182	.163	.093
14	.870	.758	.577	.442	.340	.263	.205	.160	.141	.078
15	.861	.743	.555	.417	.315	.239	.183	.140	.123	.065
16	.853	.728	.534	.394	.292	.218	.163	.123	.107	.054
17	.844	.714	.513	.371	.270	.198	.146	.108	.093	.045
18	.836	.700	.494	.350	.250	.180	.130	.095	.081	.038
19	.828	.686	.475	.331	.232	.164	.116	.083	.070	.031
20	.820	.673	.456	.312	.215	.149	.104	.073	.061	.026
22	.803	.647	.422	.278	.184	.123	.083	.056	.046	.018
24	.788	.622	.390	.247	.158	.102	.066	.043	.035	.013
26	.772	.598	.361	.220	.135	.084	.053	.033	.026	.009
28	.757	.574	.333	.196	.116	.069	.042	.026	.020	.006
30	.742	.552	.308	.174	.099	.057	.033	.020	.015	.004
40	.672	.453	.208	.097	.046	.022	.011	.005	.004	.001
50	.608	.372	.141	.054	.021	.009	.003	.001	.001	.000

Figure 95: Present value table

Two points regarding present value are apparent from the table:

1. Present value decreases as the number of years into the future increases.

   For example, using an interest rate of 10 percent, the present value of an investment worth one dollar in years 1 through 10 is shown below.

Year	*Present Value at 10%*
1	.909
2	.826
3	.751
4	.683

5	.621
6	.564
7	.513
8	.467
9	.424
10	.386

Based on an interest rate of 10 percent, a dollar received ten years from now is worth only $.39 today.

This illustrates another reason why the anticipation of a long economic life for a proposed software system is not worthwhile: with each year of projection into the future, the present value of expected benefits decreases.

2. Present value decreases as the discount rate increases.

Present values of one dollar to be received one year from now are shown below for interest rates of 4, 6, 8, 10 and 12 percent.

Year	4%	6%	8%	10%	12%
1	.962	.943	.926	.909	.893

Following Anthony and Welsch [1] present value is defined as follows: the present value (P) of an amount (F) that is expected to be received at a specified time (n) in the future is the amount, which if it were invested today at a specified interest rate (i) would accumulate to the future amount (F). The present value table shown in Figure 95 allows for easy calculation of any future amount. For example, if one expects to receive $20,000 two years from now, given an interest rate of 8 percent, the present value of the investment is calculated by multiplying $20,000 by 0.857 to arrive at a present value of $17,140.

9.3.2.2 Required Rate of Return

The present value of an amount varies, depending on the discount rate used. How does one know which rate is appropriate? The appropriate discount rate is called the *required rate of return.*

A thorough treatment of this topic is beyond the scope of this text, nevertheless, a few remarks are in order. When a bank lends funds at a certain interest rate, for example at 10 percent, it does so in the belief that repayment of the loan plus interest is almost certain. Many of the loans made by banks and other institutions are secured by land, inventory or other assets of the company. The lower interest rate demanded in these cases reflects the fact that the loans are secured and thus less risky from the lending institution's point of view.

When a company invests in a proposed software system, a new product line or equipment, however, the risk is higher and consequently the return is much less certain. The risk is higher because one cannot be certain of the benefits, or for that matter the costs!

Because of the higher uncertainty involved in ventures such as software systems and new product lines, companies generally use a higher rate of interest for the required rate of return than the current bank rate of borrowing. One often hears of business speculators who invest in speculative or risky ventures whose return on investment is very high. So it is with a proposed system: the required rate of return will be somewhat higher than the current cost of borrowing because of the risk involved.

In companies, financial officers are responsible for determining the required rate of return, not the systems analyst. For the discussions that follow, the required rate of return will be provided.

9.3.2.3 Net Present Value

We still have not arrived at a point whereby an informal judgment regarding a potential investment can be made. In this section the concepts which will form the basis for judgement are introduced.

Consider the following example. Suppose a proposed investment of $1,000 is expected to provide benefits of $500 per year for each of the next three years. The required rate of return is 10 percent. The present value of these future benefits is calculated as follows:

Year	*Benefit*	*Present Value of $1 at 10%*	*Present Value of Benefit*
1	500	.909	455
2	500	.826	413
3	500	.751	376

Present value of benefits for 3 years			1244.

However, we have not considered the costs involved with investing. These should be deducted from the benefits to arrive at a more accurate picture of an investment's worth. *Net present value* is the difference between the present value of future benefits and the cost of the investment. In our case the investment cost $1,000, so its net present value is $1,244 less $1,000, which comes to $244.

Any investment having a net present value of zero or greater is profitable. The example, then, is a worthwhile investment.

The net present value of an investment is a function of

- the amount of the investment
- the required rate of return
- the benefits it provides.

If the previous example is modified so that its economic life is changed to two years from three, the net present value is calculated as follows:

Present value of benefits for 2 years =	455 + 413 = 867
Net present value =	867 - 1000 = -$133.

The investment is no longer profitable, given the shortened economic life. Again, returning to the original example, suppose the investment is decidedly risky, and a rate of return of 24 percent is required. Our calculations are as follows:

Year	*Benefit*	*Present Value of $1 at 10%*	*Present Value of Benefit*
1	500	.806	403
2	500	.650	325
3	500	.524	262

Present value of benefits for 3 years			990
less investment			1000

Net present value			-$10

The net present value of the investment is now -$10 and again it is not profitable.

Knowing how to calculate net present value, we are in a position to compare potential investments, for the higher the net present value, the better the investment. As an example, consider two investments shown below.

	Project X	Project Y
investment	$75,000	$100,000
annual benefit	40,000	50,000
economic life	4 years	3 years
required rate of return	10%	10%

The net present value for each project is calculated below.

	Project X		Project Y	
Year	*Benefit*	*Present Value*	*Benefit*	*Present Value*
1	40,000	36,360	50,000	45,450
2	40,000	33,040	50,000	41,300
3	40,000	30,040	50,000	37,550
4	40,000	27,320		
		---------		---------
		126,760		124,300
less investment		75,000		100,000
		---------		---------
Net present value		51,760		24,300

Therefore, Project X is the more desirable of the two investments.

The examples in this section have assumed that the benefits are a constant amount each year. For Projects X and Y, the annual benefits are $40,000 and $50,000, respectively. Benefits that are a constant amount each year are called *annuities*, and the present value table can be modified for use as an annuity table, as shown in Figure 96.

Years	1%	2%	4%	6%	8%	10%	12%	14%	15%	20%
1	.990	.980	.962	.943	.926	.909	.893	.877	.870	.833
2	1.970	1.942	1.886	1.833	1.783	1.736	1.690	1.647	1.626	1.528
3	2.941	2.884	2.775	2.673	2.577	2.487	2.402	2.322	2.283	2.106
4	3.902	3.808	3.630	3.465	3.312	3.170	3.037	2.914	2.855	2.789
5	4.853	4.713	4.452	4.212	3.993	3.791	3.605	3.433	3.352	2.991
6	5.795	5.601	5.242	4.917	4.623	4.355	4.111	3.889	3.784	3.326
7	6.728	6.472	6.002	5.582	5.206	4.868	4.564	4.288	4.160	3.605
8	7.652	7.325	6.733	6.210	5.747	5.335	4.968	4.639	4.487	3.837
9	8.566	8.612	7.435	6.802	6.247	5.759	5.328	4.946	4.772	4.031
10	9.471	8.983	8.111	7.360	6.710	6.145	5.650	5.216	5.019	4.192
11	10.368	9.787	8.760	7.887	7.139	6.495	5.937	5.453	5.234	4.327
12	11.255	10.575	9.385	8.384	7.536	6.814	6.194	5.660	5.421	4.439
13	12.134	11.343	9.986	8.853	7.904	7.103	6.424	5.842	5.583	4.533
14	13.004	12.106	10.563	9.295	8.244	7.367	6.628	6.002	5.724	4.611
15	13.865	12.849	11.118	9.712	8.559	7.606	6.811	6.142	5.847	4.675
16	14.718	13.578	11.652	10.106	8.851	7.824	6.974	6.265	5.954	4.730
17	15.562	14.292	12.166	10.477	9.122	8.022	7.120	6.373	6.047	4.775
18	16.398	14.992	12.659	10.828	9.372	8.201	7.250	6.467	6.128	4.812
19	17.226	15.678	13.134	11.158	9.604	8.365	7.366	6.550	6.198	4.844
20	18.046	16.351	13.590	11.470	9.818	8.514	7.469	6.623	6.259	4.870
22	19.660	17.658	14.451	12.042	10.201	8.772	7.645	6.743	6.359	4.909
24	21.243	18.914	15.247	12.550	10.529	8.985	7.784	6.835	6.434	4.937
26	22.795	20.121	15.983	13.003	10.810	9.161	7.896	6.906	6.491	4.956
28	24.316	21.281	16.653	13.406	11.051	9.307	7.984	6.961	6.534	4.970
30	25.808	22.396	17.292	13.765	11.258	9.427	8.055	7.003	6.566	4.979
40	32.835	27.355	19.793	15.046	11.295	9.779	8.244	7.105	6.642	4.997
50	39.196	31.424	21.482	15.762	12.234	9.915	8.304	7.133	6.661	4.999

Figure 96: Annuity table

Annuity tables give the cumulative present values for a range of economic lifetimes and rates of return. The values that appear in annuity tables are based on an investment that is worth one dollar at the time it is n years old.

For Project X in the previous example, the present value of the benefits is calculated using a 10 percent rate of return by

$$40{,}000 \times 3.170 = \$126{,}760$$

The figure 3.170 is selected from the annuity table because the economic life of the investment is four years and the required rate of return is 10 percent. This figure can also be derived by referring to the present value table in Figure 95 and adding four years of present values in the 10 percent column. (The results have been rounded.)

$$0.909 + 0.826 + 0.751 + 0.683 = 3.169$$

The discussion of net present value concludes with three remarks. First of all, the calculation of an investment's net present value is an estimation of several things: the cost of the investment, the required rate of return and the anticipated benefits of the investment. Of these three, only the required rate of return is certain; the anticipated costs and benefits are uncertain. Therefore, when calculating the present value, the results are rounded to whole dollars. In fact, in most realistic investment scenarios, the results are rounded even further, usually to the thousands of dollars. To show portions of dollars gives a false impression of accuracy.

Secondly, the present value and annuity tables in this chapter are constructed on the assumption that the benefits are received once a year, on the last day of the year (starting with today as the beginning of the year). For a proposed system (and for other investments such as new product lines or a new branch office) this is not realistic because the benefits are expected to occur throughout the year. However, these annual tables are normally used in business investment problems because

- They are easier to understand than tables constructed on the assumption that benefits will occur continuously or monthly throughout the year.

- They provide estimates that are close enough to warrant their use.

Finally, for a proposed system, the cost of the investment is its development cost. Therefore, the net present value of a proposed system is the present value of the benefits minus the development costs.

9.3.3 Estimating the Net Present Value of a Proposed System

All of the concepts related to estimating the economic worth of a proposed software system have now been discussed: development costs, operating costs, benefits, economic life, required rate of return, present value and net present value.

The discussion now focuses on isolating and estimating the elements involved in such calculations, and how these elements fit together to give the economic assessment of a proposed system.

In order to estimate the net present value of a proposed system, one must be able to estimate or be given the following cost/benefit estimates:

1. the economic life of the proposed system

2. the required rate of return

3. the development costs of the proposed system

4. the annual operating costs of the current system (This means the costs pertaining to the processes that will change with the installation of the proposed system [7].)

5. the annual operating costs of the proposed system

6. the annual benefits anticipated from the proposed system.

The derivation of each item is discussed by means of an example.

9.3.3.1 Estimating Economic Life, Required Rate of Return and Development Costs

Company management, user management and data processing management, together with the analyst, should provide an intelligent estimate of the proposed system's economic life. For discussion purposes, an economic life of five years will be used. As well, suppose that the financial officers have indicated that the required rate of return is 16 percent.

To estimate the development costs of the proposed system, the analyst may find it useful to refer to previous systems development projects undertaken by the company. Taking a look at past projects may provide some insight into the associated cost categories and the methods used to derive past estimates.

Development costs usually fall into one of five categories.

- Personnel
- Equipment
- Materials and Supplies
- External Costs
- Overhead

A checklist of typical development cost elements appears at the end of this chapter. Recall that personnel costs involve more than just salaries; they include company benefits also. The costs of company benefits are usually estimated as a percentage of salaries. For example, if it is estimated that an analyst whose current salary is $40,000 will work one year starting now on the project, and if personnel benefits are estimated to cost the company 30 percent, then the cost to the company for the analyst's participation in the project is

$$\$40{,}000 \times 1.3 = \$52{,}000.$$

Major costs in each category are estimated and totalled to give an estimate of the total development cost.

Suppose the following development considerations have been identified. Two years of development are required, starting next year (years 1 and 2). Major costs have been isolated as follows:

Personnel

- Company benefits are estimated at 30 percent of salary.
- One analyst will work one full year in year 1. The salary estimate for next year is $40,000.
- Four programmers will be assigned to the project for year 2. The average salary estimate for year 2 is $32,000.

- Other development personnel costs (for computer operators, clerical personnel, management supervision and so on) are estimated at $18,000 in each of years 1 and 2.

Personnel costs can be summarized as follows:

	Year 1	*Year 2*
Development Costs:		
Analyst	40,000	
Programmers		128,000
Other	18,000	18,000
Benefits (30%)	17,400	43,800
	---------	---------
Total Personnel costs	75,400	189,800

Other Development Costs

- The proposed system will operate on existing equipment. However, the purchase of an additional disk drive will be required for the system. It will be purchased in year 2. The estimated cost is $30,000. The estimated cost of installation and testing is $3,000.

- Testing of programs and the system, and converting to the new system will occur in year 2. The estimated cost is $90,000.

- Materials and supplies are estimated to cost $5,000 in year 1 and $10,000 in year 2.

- Overhead is estimated to be 5 percent of development costs.

- Because no new software or hardware (other than the new disk drive) is being purchased, there are no external development costs.

These development costs are summarized below:

	Year 1	Year 2
Development Costs:		
Equipment Purchase		30,000
Equipment Testing & Installation		3,000
System Testing & Conversion		90,000
Materials & Supplies	5,000	10,000
Overhead - 5%	250	6,650
	---------	---------
	5,250	139,650

All development costs from the previous two tables are summarized below.

	Year 1	Year 2
Development Costs:		
Personnel	75,400	189,800
Other	5,250	139,650
	---------	----------
Total Development Costs	80,650	329,450

9.3.3.2 Estimating Annual Operating Costs of the Current System

Management responsible for the area under study will be able to provide the analyst with estimates of operating costs for the current system. Only those costs that will be eliminated by the implementation of the proposal need be isolated. Next, these costs are projected over the economic life of the new system. This is done because the elimination of these costs is a benefit (cost savings) provided by the new system.

Suppose the following operations will cease when the proposed system is operational:

- Four data entry personnel will be eliminated; their current average salary is $20,000. Salaries are expected to increase by 5 percent per year for the foreseeable future.

- Equipment used by the data entry personnel will also be eliminated. This equipment is currently leased at an annual cost of $24,000. It is anticipated that lease costs will increase by 8 percent per year.

- Other associated costs that will cease, such as supervision, supplies and overhead are estimated to be $18,000 per year. It is expected that these costs will increase by 10 percent per year.

This information is summarized below in table form. Notice that the present is designated as Year 0.

Year	0 (now)	1	2	3	4	5	6	7
Current System Operating Costs:								
Personnel	80,000	84,000	88,200	92,610	97,241	102,103	107,208	112,568
Benefits	24,000	25,200	26,460	27,783	29,172	30,631	32,162	33,770
Equipment Lease	24,000	25,920	27,994	30,234	32,653	35,265	38,086	41,133
Other	18,000	19,800	21,780	23,958	26,354	28,989	31,888	35,077
Total	146,000	154,920	164,434	174,585	185,420	196,988	209,344	222,548

Because the cost estimates are subject to yearly increases, they are projected through the years of development (years 1 and 2). These current system operating costs will not be eliminated, however, until the proposed system becomes operational in year 3. The economic life of the system is five years, so we project the costs through years 3 to 7.

9.3.3.3 Estimating Annual Operating Costs of the Proposed System

The annual costs of operating the proposed system must be determined. A checklist of typical operating costs for systems appears in section 9.6.2 at the end of this chapter. Of course, not all costs apply to every situation.

The analyst can be aided in the task of estimating annual operating costs of a proposed system by an examination of previous development projects and by discussions with data processing management and user management.

Suppose that for the proposed system the following operating costs have been isolated:

- Computer resource usage for the proposed system is expected to remain about the same as it has been for the current system, although the amount of available disk space will be increased. The computer costs associated with the increased disk space are estimated to be $20,000 during the first year of operation, with a 5 percent increase in each subsequent year.

- Personnel costs for operator and clerical support, maintenance programmers and management supervision are expected to be similar to personnel costs for the current system (less the data entry functions that will be eliminated).

- Data entry services will be performed external to the company by a local service bureau and the results will be transmitted to the computer site. Estimated costs for this service are $40,000 during the first year of operation, with an expected 10 percent increase in each subsequent year.

- The costs pertaining to the operation of the additional disk drive include the increased maintenance costs of hardware and the purchase and maintenance of additional removable disk packs. These costs are estimated to be $5,000 during the first year of operation with an expected increase of 5 percent for each year following.

These costs are projected over the life of the system and are shown below.

Year	3	4	5	6	7
Proposed System Operating Costs:					
Computer	20,000	21,000	22,050	23,153	24,311
External	40,000	44,000	48,400	53,240	58,534
Materials & Maintenance	5,000	5,350	5,513	5,789	6,078
Total	65,000	70,350	75,963	82,182	88,923

9.3.3.4 Estimating Annual Intangible Benefits of the Proposed System

We are now in a position to tackle the most difficult aspect of the quantification of the proposed system: how are intangible benefits such as faster processing of customer orders, improved position in the market place, more accurate management information, or an improved work environment to be quantified? In order to calculate and compare benefits, a monetary value must be placed on them.

Some of these intangible benefits are easier to quantify than others. However, in all cases, the systems analyst must never be the one to attempt to quantify them. Because the user will benefit by the proposed system, the user should be the main source of quantification. Company auditors and financial officers may be able to assist the user in deriving such estimates.

To continue with the example developed thus far, suppose it is anticipated that the proposed system will provide the following benefits:

- Customer orders, invoices and accounts will be processed more rapidly. Currently accounts receivables total $1,200,000 and are expected to continue at this level.

- It is estimated that the faster processing of customer orders and invoices will mean that the time over which receivables remain outstanding will decrease. This will result in a 10 percent reduction of receivables in each year that the new system is operational. This reduction (increased cash to the company) will be applied to a bank loan that the company currently has at the bank. The annual rate of interest of this loan is 15 percent.

The accounts receivable reduction is shown below.

Year	2	3	4	5	6	7
A/R	1,200,000	1,080,000	972,000	875,000	787,000	708,000
A/R reduction		120,000	108,000	97,000	88,000	79,000

The savings will be applied in the form of payments to the bank loan. The benefit to the company is reduced bank interest. The savings in interest is as follows:

YEAR	INTEREST
3	120,000 x .15 = 18,000
4	108,000 x .15 = 16,200
5	97,000 x .15 = 14,550
6	88,000 x .15 = 13,200
7	79,000 x .15 = 11,850

Because these payments would occur throughout the year, we estimate that half of the interest will be saved in each year. The benefit then becomes

Year 3	*Year 4*	*Year 5*	*Year 6*	*Year 7*
9,000	8,100	7,275	6,600	5,925

Another anticipated benefit is the provision of better information to the sales force, until now, a problem with the current system. The result of providing poor information has been a large turnover in salespeople, at considerable expense to the company. By providing better information to salespeople, and giving them more support from management, is it hoped that more salespeople will remain, thereby reducing the costs associated with employee turnover.

Suppose sales management estimates that sales force turnover will be eight people per year in each year of operation of the new system. Currently thirteen salespeople per year leave the company. The benefit this provides over the life of the system is the annual savings of five people not leaving the company. The personnel department has supplied the following costs relevant to one salesperson leaving the company.

Severence pay, etc.	$4,000
Cost to hire replacement:	
- advertising	1,000
- recruiting, interviewing, etc.	2,000
Cost of new employee:	
- Training classes	2,000
- Salary and benefits during training	20,000

	$29,000

These are current costs which are expected to increase by 5 percent each year. These benefits are projected through years 0 to 7 and are shown below.

($)

Year	0	1	2	3	4	5	6	7
Cost of 5 Salespeople Leaving	145,000	152,250	159,863	167,856	176,249	185,061	194,314	204,030

Total intangible benefits can be summarized as follows:

Year	3	4	5	6	7
Savings in Bank Interest	9,000	8,100	7,275	6,680	5,925
Savings in Personnel Costs	167,856	176,249	185,061	194,314	204,030
Total	176,856	184,349	192,336	200,994	209,955

All the cost and benefit elements can be assembled in order to calculate the present value of the proposed system.

Year	3	4	5	6	7
Current System Operating Costs (less)	174,585	185,420	196,988	209,344	222,548
Proposed System Operating Costs	65,000	70,250	75,963	81,182	88,923
Operating Benefit Provided by New System	109,585	115,170	121,025	128,162	133,625
Intangible Benefits	176,856	184,349	192,336	200,994	209,955
Total Benefits	286,441	299,519	313,361	329,156	343,580

The results are converted to their present values, given a required rate of return of 16 percent.

Year	Benefit	Present Value of $1 at 16%	Present Value of Benefit
3	286,441	.641	183,609
4	299,519	.552	165,334
5	313,361	.476	149,160
6	329,156	.410	134,954
7	343,580	.354	121,627

			754,684

The present value of the proposed system is $754,684.

To calculate the net present value, development costs are subtracted. Development costs from section 9.3.3.1 are as follows:

Year	Benefit	Present Value of $1 at 16%	Present Value of Benefit
1	80,650	.862	69,520
2	329,450	.743	244,781

			314,301

Thus the net present value of the proposed system is

$754,684 - 314,301 = $440,383

so the system is economically feasible.

9.4 Cost/Benefit Analysis - Summary

To a company, a proposed software system represents an investment opportunity, but it is just one of the many opportunities that competes for available capital. In order to assess the particular opportunity offered by a new software system, management must be provided with estimates of all anticipated costs and benefits. The procedure described in this chapter for isolating and estimating these considerations – a cost/benefit analysis – is performed by the analyst with the help of certain areas of management within the company: user, data processing and financial management.

A cost/benefit analysis of a proposed system has the following components:

- a cash flow analysis (table) of the proposed system for its development period and economic life
- the determination of the proposal's present value and net present value.

The steps necessary for this procedure are

1. estimation of the proposed system's economic life
2. determination of an appropriate required rate of return
3. estimation of the development costs for the proposed system
4. estimation of the annual operating costs of the processes that will be eliminated with the implementation of the proposal
5. estimation of the annual operating costs of the proposed system
6. estimation of the annual benefits anticipated with the proposed system.

The difference between the present system's annual operating cost (4), and the annual operating costs of the proposed system (5) is sometimes referred to as the *operating benefit* provided by the proposed system. This implies that the new system will cost less to operate than the present system. However, there are cases where this is a negative benefit (a cost). When this occurs, the anticipated benefits (6) should offset this cost.

After estimates for each of the six items have been obtained and projected over the economic life of the proposed system for each year of operation, several calculations are made.

- The operating costs of the present system are subtracted from the annual operating costs anticipated for the proposed system.

- The results (either negative or positive) for each year are added to the estimates of the annual benefits expected. The result is an estimate of net benefits for each year of operation.

- Each year's benefit is then converted to its present value, and all are summed.

- Development costs for the proposed system are converted to their present value and subtracted from the present value derived from the above step.

The result is the net present value, or economic assessment, of the proposed system. The analyst must provide a cash flow analysis to accompany the net present value assessment of the proposed system. All the calculations required for a cash flow analysis have been performed, for the entries in a cash flow table (Figure 93) are the estimates of annual costs and receipts projected over the development and operational phases of the system (years 1 through 7). The cash flow table shown in Figure 93 has row labels for each of the following categories:

- Beg. Bal. - beginning balance. At the beginning of the project, no costs have been incurred, so the balance is zero.

- Devel. Costs - development costs. Development costs for each year are indicated.

- Operating Costs - operating costs of the proposed system.

- Receipts - the operating costs of the current system (the costs that will cease when the new system becomes operational) plus intangible benefits.

- Ending Bal. - ending balance.

Ending Balance = Beginning Balance - Development Costs - Operating Costs + Receipts

Notice that in Figure 93 negative amounts are parenthesized.

The cash flow table shown in Figure 93 gives annual breakdown of cash flow. Cash flow tables may also give quarterly or semi-annual breakdowns.

Anticipated intangible or *non-operating* benefits are difficult to quantify, but if they are expected to provide a return to the company, then an estimation of their benefit must be undertaken. Often for proposed systems, management anticipate an increase in sales or some other productivity gain that will result in increased profits. If an anticipated benefit is of this nature, then the benefit is what is realized **after taxes**. For example, suppose we expect to receive a gain of 8 percent in sales because of our new system. Assume that the company's history has been a net income of 10% after taxes. If current sales are $1,000,000, then the net benefit is

$$(.08)(\$1{,}000{,}000)(.10) = \$8{,}000$$

because only 10 percent of the increase in sales is realized as income.

9.5 Establishing a Business Case

Once cost and benefit estimates for various alternative approaches have been made and a recommended final approach decided upon, detailed costs and benefits for the chosen alternative must be developed. This information forms a part of the business case for the recommended system. The business case basically summarizes what corporate needs necessitated the project, what was discovered, what automation concept was created and what is recommended and why. All of this material is written into a document normally called the business case and accompanied by the details of the results of system analysis, the functional specification.

The typical table of contents of a business case will be as follows:

Table of Contents

I. Executive Summary
(a one-page summary of objectives, scope, reason for the analysis and recommended solution)

II. Introduction and Background
(business reasons for the analysis, a summary of the requirement specification).

III. Ideal System (overview)

IV. Selected Solution
(overview of options/strategies for automation, rationale for selected solution and make-buy decision)

V. Detailed Cost/Benefit Analysis

VI. Implementation Plan
VII. Summary and Conclusions
Appendix: Data dictionary entries for all data and process logic, DFD's, E-R models)

Once this document is prepared, its contents are presented to user and systems management for both their edification and approval. This presentation is often a formal one, with audio-visual aids.

Finally, management (usually, but not always, the user management) must use this document to make a business case and win corporate approval for obtaining the human and financial resources to implement the system.

9.6 Typical Cost Elements

9.6.1 Development Cost Categories

- *PERSONNEL*
  - Analysts
    - Interviewing
    - Preparation of reports, documentation, presentations, procedures, etc.
    - Testing
    - Training users, operators, clerical staff
    - Supervision
    - Presentations
    - Meetings
  - Programmers

  - Coding
  - Documentation
  - Debugging
  - Inspections, walkthroughs
  - Meetings
- Operators
  - Training
  - Conversion
  - Meetings
- Clerical Personnel
  - Training
  - Conversion
  - Meetings
- Management
  - Supervision
  - Meetings
- Other
  - Data entry
  - Secretarial

- *EQUIPMENT*
  - New equipment
  - Packaged software
  - Equipment installation, testing, debugging
  - Existing equipment use - programming, testing and conversion phases
    - CPU time
    - Disk space used
    - Number of I/O transfers
    - Number of output lines printed
    - Supplies - tapes, disks, etc.
- *SUPPLIES AND MATERIALS*
  - Procedures manuals
  - Stock tab paper, special forms
  - Photocopies
- *OVERHEAD*
  - Management support
  - Secretarial support
  - Utilities, space
- *EXTERNAL*
  - Consultants' fees
  - Auditors' fees
  - Training costs

9.6.2 Operating Cost Categories

- *HARDWARE*
  - CPU time
  - Disk space used
  - Number of I/O transfers
  - Number of output lines produced
  - Communications costs
  - Supplies - tapes, disks, etc.
  - Tape storage costs
  - Maintenance
- *PERSONNEL*
  - Secretarial support
  - Maintenance programmers
  - Operator support
  - Management support
  - Data entry support
- *OVERHEAD*
  - Management support
  - Secretarial support
  - Utilities, space

- *EXTERNAL*
- Leases and rentals
- Contracting

9.7 Exercises

1. Do you agree or disagree with the statement that "information systems projects are really research and development?" What are the similarities between R & D and systems development? What are the differences?

2. Operational feasibility refers to the compatibility of a proposed system to current company procedures. Illustrations of systems that are operationally unfeasible include unopened computer reports and unused terminals in executives' offices. What is the relationship between operational feasibility and economic feasibility? Which depends on the other?

3. Why, when considering the long term, are all costs variable? Give several examples of costs that may be fixed for a period of time but are ultimately variable.

4. Why are the salaries of the analysts, programmers and other development staff considered to be differential costs?

5. Why are the computer resources required by a software system considered in a cost/benefit analysis?

6. Draw a graph of an investment over its economic life. Illustrate the probable decline in returns as it nears the end.

7. Develop a cash flow table for the example developed in section 9.3.3 of this chapter.

8. Suppose that one of the aspects of a proposed system will be the upgrading of skills for direct users. Should this be considered a benefit in a cost/benefit analysis? Why?

9. Consider the following information regarding a proposed software system.

Current System
Four employees whose current average salary including benefits is $20,000 operate the system. Equipment operating costs for the equipment they use is $10,000. The special forms and other supplies for the equipment cost $8,000.

Proposed System
The new system will eliminate the data entry functions described above but one employee will remain to operate the new system. The equipment currently used (described above) is obsolete, and is being scrapped. The new system will be installed at the end of next year, and should last for five years. To replace this data entry function, the company will purchase time-sharing at a local company. This local company currently charges the flat rate of $100 per hour for computer time plus $20.00 per hour labor for data entry. It is estimated that the new system will require 25 computer hours per year and 1000 hours of data entry time. (The results of data entry will be transmitted electronically to the new system, and the associated costs built into the computer time charge.)

The computer resources required by the new system are estimated to be the same as for the present system.

Two analysts, now working at $40,000 each (including benefits) will work on the system starting at the beginning of next year. The project will require 1.5 person-years of analysis. Two junior programmers now working at $22,000 will work next year from July to December on the project. Training costs for personnel will probably be $5000 for the year of development and $1000 for each of the following two years. In-house testing (today's rate, $200 per hour) will require 50 hours of computer time next year and 10 hours the year after. It is hoped the system will return intangible benefits of $20,000 in the first year, with an annual increase of 10 percent. The required rate of return is 15 percent, and it is assumed all costs will increase by 5 percent per year.

a. Create a cash flow table for the proposed system.

b. What is the present value of the proposed system's benefits?

c. Considering the annual benefit to be the average benefit over the life of the system, what is the payback period?

10. A rubber company manufactures tires and other items made of rubber. The processing of orders is currently a manual process. Many of the company's customers have retail stores and place their orders by telephone. It is anticipated that customer service will be improved by allowing customers to access the pending order database directly, were the system automated. As part of analyzing the order processing function, a complete cost/benefit study is to be undertaken.

    It is expected that an automated system would last five years, and management hopes to install the system two years from now. The total development cost is expected to be $250,000, occurring in equal amounts during each of the next two years.

    The company estimates that each order currently costs $29.00 to process, from its receipt through to the final shipment. Marketing has projected that the sales volume will reach 2 million orders by the end of this year, and will continue to increase by 10 percent each year for the next eight years. Analysts working on the new system model have projected that using the new system, each order will cost the company $18.00 to process. As well, it is hoped that the new system will improve the company's position in the marketplace. This should result in an additional benefit of $0.20 savings per order during the first year of operation with an additional $0.05 savings per order in each subsequent year. The financial officer has indicated that the required rate of return is 15 percent.

    a. What is the present value of the proposed system?

    b. Develop a cash flow table for the system.

11. Suppose you have assembled the following information regarding a proposed software system.

    - The proposed system is to be installed three years from now, and its projected economic life is six years.

    - The operating costs of the present system are estimated to be $450,000 for next year.

    - The operating costs of the proposed system are estimated to be $300,000 during the first year of its operation.

    - The value of intangible benefits of the proposed system is projected to be $180,000 during the first year of its operation.

- The operating costs of the present system, the operating costs of the proposed system, and all benefits are expected to increase by 10 percent per year.

- The required rate of return is 16 percent.

- The development costs of the proposed system are expected to be $650,000; $250,000 will be spent on development next year, and $200,000 will be spent on development during each of the subsequent two years.

a. Calculate the net present value of the proposed system. Is the system economically feasible? Why?

b. Develop a cash flow table for the proposed system.

9.8 Bibliography

1. Anthony, Robert N. and Welsch, Glenn A. *Fundamentals of Management Accounting.* Homewood, Illinois. Richard D. Irwin, 1977.

2. Biggs, Charles L., Birks, Evan G. and Atkins, William. *Managing the Systems Development Process.* Englewood Cliffs, New Jersey. Touche Ross and Company, Prentice-Hall, 1980.

3. Cash, James I., Jr., McFarlan, F. Warren and McKenney, James L. *Corporate Information Systems Management: Text and Cases.* Homewood, Illinois. Richard D. Irwin, 1983.

4. Curtis, Bill (editor). *Tutorial: Human Factors in Software Development.* Los Angeles, California. IEEE Computer Society, 1981.

5. Dampney, C.N.G. and others. *Data (Information) Analysis of Business Systems - Informal Notes.* Waterloo, Ontario. University of Waterloo, 1986.

6. DeMarco, Tom. *Structured Analysis and System Specification.* Englewood Cliffs, New Jersey. Prentice-Hall, 1979.

7. Gildersleeve, Thomas R. *Successful Data Processing System Analysis, Second Edition.* Englewood Cliffs, New Jersey. Prentice-Hall, 1985.

8. Horngren, Charles. *Cost Accounting: A Managerial Emphasis, Fifth Edition.* Englewood Cliffs, New Jersey. Prentice-Hall, 1985.

9. Kleijnen, Jack P.C. *Computers and Profits: Quantifying Financial Benefits of Information.* Reading, Massachusetts. Addison-Wesley, 1980.

10. McFarlan, F. Warren and McKenny, James L. *Corporate Information Systems Management: The Issues Facing Senior Executives.* Homewood, Illinois. Richard D. Irwin, 1983.

11. Nolan, R.L. (editor). *Managing the Data Resource Function.* San Francisco, California. Benjamin West, 1982.

12. Parkin, A. *Systems Management.* London, U.K. Edward Arnold, 1980.

13. Pressman, Roger S. *Software Engineering: A Practitioner's Approach.* New York, New York. McGraw-Hill, 1982.

14. Rockart, John F. and Bullen, Christine V. (editors). *The Rise of Managerial Computing: the Best of the Center for Information Systems Research.* Homewood, Illinois. Dow Jones-Irwin, 1986.

15. Sprague, Ralph and McNurlin, Barbara. *Information Systems Management in Practice.* Englewood Cliffs, New Jersey. Prentice-Hall, 1986.

Part 3

Fourth-Generation Technologies and Systems Prototyping

Chapter 10

SHORTCOMINGS OF THE TRADITIONAL APPROACH

The tools and methods of systems development presented in Part 2 represent the traditional approach employed in most corporations today. We shall now discuss certain problems occurring in the data processing industry that are associated with the traditional approach and discuss alternative approaches that have been developed recently which are designed to alleviate problems which the traditional approach cannot address. The terms traditional approach, life cycle and traditional life cycle all mean the same thing and are used interchangeably.

10.1 The Creation of the Traditional Life Cycle

The generic life cycle presented in Chapter 2 is typical of the software development process followed in most major corporations. It was developed to solve some of the severe problems encountered in the data processing industry during the 1960's. Symptoms of these problems were:

- Late projects
- Cost overruns
- Systems not meeting requirements
- Very long development times

- Error-prone systems (infested with bugs or unusable)
- Very difficult and expensive maintenance
- Inflexible systems
- Systems incompatible with current procedures

The underlying problems causing these symptoms were traced to the following:

- Lack of understanding of business requirements by systems developers
- Difficulty in estimating programming effort
- Developing programs in low-level languages
- Poor-quality programs (unreadable, error prone)
- Lack of design and programming standards
- Poor communication between users and data processing personnel
- Increasing complexity of applications

The life cycle was created to approach development in a systematic and consistent fashion, much like the process followed in manufacturing an automobile. Requirements were to be totally compiled and specified in the first phase, a system conceived in the second phase, designed in the third, programmed in the fourth, tested and implemented in the last two phases. During this process, documentation was to be maintained on all aspects of the system. Structured programming was to make the resulting code readable, correct and maintainable. Tools were developed (e.g. data flow diagrams, structured English) for communication between users and developers. The life cycle, then, was an attempt to lay out a series of integrated steps leading from requirements to an "engineered" system.

The traditional approach has undoubtedly meant the difference between successful systems development and disastrous failures. In fact before its inception there *were* failures, typically development efforts which took so long or were so complex that nobody knew what was going on. The strengths of the life cycle are that it requires careful thinking, specification and planning, especially in the first phases. In fact for certain types of applications requiring what we might call "tight" design, one cannot envision doing things in any other way. For example, the design of real-time reservation or banking systems needs to be very precise to handle specific transactions and to achieve fast responses.

10.2 The Changing Nature of Applications

The earliest business computer systems were designed to process large volumes of data, thereby replacing very repetitive manual procedures. A typical example is the payroll of a large company. Employee hours are input, deductions are computed, checks printed and reports generated. Very few changes are ever made to this system because it has a single, well-defined purpose. Such systems are in wide use and are often referred to as production or operational support systems.

In the past few years new types of applications have become more prevalent. These include decision support systems and so-called ad hoc applications, among others. The newer types of applications are required by middle and senior managers to help them extract information out of the mass of data produced by the operational systems or from external computer systems.

A decision support system allows managers to extract information from a corporate or other computer and to analyze one or more situations based on this information. For example a manager could evaluate the impact of different salary increases on the company's budget, or predict the company's profits if prices rise but sales decrease. The point is that a flexible system is required, which means that managers cannot possibly predict in advance the scenarios to be evaluated or even what information will be required. What managers really need is a system that allows them to easily create their own applications, or at least one that allows someone else to build an application quickly.

The ad hoc type of application is one in which certain information is required immediately, and if it is not obtained quickly, it is worthless and an opportunity has been missed. This type of application is often needed only once. For example, suppose that a marketing manager plans to meet with a major client to discuss an exclusivity agreement based on the volume of business between the companies. Figures showing past business with the client will certainly be needed, but if current reports do not compute volume by client by year, an ad hoc application will need to be created to extract this information. Again, the manager may not be able to predict in advance what is desired on the report, and may not even know the report is needed until a few days before meeting with the client.

10.3 Application Backlogs and Maintenance

In all too many corporations the current state of data processing can be characterized by three- to five-year backlogs of application requests and by a disproportionate share of resources spent on systems maintenance. In some organizations nearly 100 percent of programming budgets are spent on maintenance.

The seriousness of the application development backlog cannot be overrated, for it is not merely a situation of disenchanted users. Rather, an unresponsive data processing environment, unable to keep pace with the dynamic business environment, will reduce the effectiveness of management. They will perceive opportunities slipping by and potential benefits lost by not being able to access new information or to develop new systems to meet their needs quickly. If they must wait two or three years for a system, it, and the assumptions behind it, will surely be obsolete. The required system is therefore less useful than if it could be delivered immediately. By the time it is ready, changes in the business environment will make it necessary to enhance the system immediately, and the user has to wait again for the system.

The traditional life cycle is not capable of handling continual changes of requirements. It must therefore evolve or be replaced to reflect the realities of the modern business environment and new developments in hardware and software.

The basic procedures followed in the life cycle (as described in Chapter 2) can be traced to the computer hardware and software for which its products were destined. Most applications are ultimately coded in what are termed third-generation procedural languages such as FORTRAN and COBOL, on third-generation hardware. This means that all of the processes or functions of each program must be known and specified in advance and to the lowest level of detail possible.

The project team must, through a series of transformations, translate business-type requirements into precise step-by-step procedures. All aspects of this transformation must be documented and kept current. Different terminology is required for each phase of the project, starting with the user's terminology, followed by technical data processing terminology and finally COBOL or FORTRAN. So, as a project proceeds, it becomes increasingly difficult for users and systems people to communicate. To further complicate matters, many applications must interface with other new or existing ones and all applications must be maintained over their lifetimes. Thus, there is a strong requirement for consistent and compatible systems development, and the life cycle was invented to fill that very requirement.

10.4 Shortcomings of the Traditional Life Cycle

In actual practice, while we do find a full range of opinions and experiences among users and systems developers, there is much general dissatisfaction with the traditional approach, and this dissatisfaction is growing. Many companies have spent large sums of money and considerable effort in establishing their software development life cycles. Often volumes of documentation exist, spelling out processes and procedures for systems development. Users and developers are often overwhelmed when faced with such a highly structured bureaucratic process. Most would agree that a major inadequacy of the traditional approach is the frustration it can create from the very beginning of a project.

Another problem with the life cycle is that it tacitly assumes that all requirements are known and can be specified to the smallest level of detail at the outset of the project. This assumption is implied by the sequential nature of the six phases. Unfortunately, several serious problems arise in practice.

First of all, in any complex business problem, it is very unlikely that all conditions or possibilities can be thought of by a team, for the team members represent a small subset of the possible user group. Also, some requirements are essential and others just "nice to have," but it is often a matter of personal opinion just where which to draw the line between what is essential and what is not. Further, there is a human tendency to give answers because they are required, even if they are not known. Consequently functions which may be asked for are not correctly defined or thought out and provided. Incorrectly reasoned out functions inevitably result in systems maintenance requests later, and they cost a great deal to remedy. Another reality is that requirements are dealt with on paper, or what we might call conceptually, that is, users are expected to specify everything in advance without ever having seen a sample of the final product. When these specifications are implemented, users often perceive the system differently and admit "that's not exactly what I had in mind." For example the layout of an input screen and its speed of execution may be different in practice from the description on paper.

Probably the most frustrating aspect of the traditional life cycle is the long lead time between requirements specification and implementation. Part of this problem can be attributed to coding in procedural languages, to the design effort this requires, and to the long time required for testing. But more fundamentally, all or most of the work is performed manually because there have been few automated aids available until recently. This is ironic, for in fact the very people who are automating others are in many cases the least automated. One may wonder at this state of affairs and reflect on why the traditional life cycle is a manual process after a decade of use. The fact is that attempts to automate it have been made and partial success achieved, but the problem is extremely complicated. Real success would require a system capable of handling complex applications from very diverse areas, yet fully general in that applications of any type can be created.

The real challenge is not so much in automating the individual phases, for example by creating requirements specification, automated data flow programs, etc, but in automating and integrating the entire process. This has not been accomplished, and indeed, it may not be worth the effort. To automate the existing life cycle may be automating the wrong thing. Instead, it appears that we must create entirely new approaches to applications development. We will discuss some current progress in this direction in Chapters 11 and 12.

The life cycle, then, has a place in modern systems development, but its inadequacies require new approaches. The remainder of this book is devoted to discussing some alternative tools and methods now being utilized in many companies.

10.5 Exercises

1. How could automation help alleviate some of the drawbacks to the traditional life cycle?

2. If design and programming were automated, how would testing be impacted? Draw a new life cycle under the assumption that design and programming are automated. Do not view software creation as a sequential process.

3. Regardless of how automated software creation becomes, some aspects of the traditional approach will always be required. What are these?

4. Many users and systems people view the traditional approach as being bureaucratic, ponderous and slow. Are they justified in this opinion? Discuss your position. Remember, there are two sides to every story.

5. Throughout Part 2 of this text, we have concentrated on the use of the data flow diagram and associated tools and methodology for systems analysis. However, many other methodologies have been developed, and different companies have either developed or modified the methodologies to meet their own needs.

   a. Investigate a systems analysis/design methology different from what has been discussed (this is also referred to as the "Gane-Sarson/DeMarco" methodology).

   b. Write a short report regarding the methodology you investigate. The report should include a description of the methodology, a comparison with the one discussed in Part 2 of the text (advantages and disadvantages) and a small example of its use, if possible. The report should be approximately ten pages long.

   Examples of other methodologies are: SADT, PSL/PSA, JSD, Warnier-Orr and Higher Order Software.

10.6 Bibliography

1. Boar, Bernard H. *Application Prototyping: a Requirements Definition Strategy for the 80s*. New York, New York. John Wiley & Sons, 1984.

2. Budde, R., Kuhlenkamp, K., Mathiassen, L. and Zullighoven, H. (editors). *Approaches to Prototyping*. Berlin, West Germany. Springer-Verlag, 1984.

3. Carey, T.T. and Mason, R.E.A. *Information System Prototyping: Techniques, Tools, and Methodologies.* INFOR, Volume 21 Number 3. August, 1983

4. Cash, James I., Jr., McFarlan, F. Warren and McKenney, James L. *Corporate Information Systems Management: Text and Cases.* Homewood, Illinois. Richard D. Irwin, 1983.

5. Joyce, Edward J. *The Art of Space Software.* DATAMATION, Volume 31, no. 22, pp. 30-34. November 15, 1985.

6. Martin, James. *An Information Systems Manifesto.* Englewood Cliffs, New Jersey. Prentice-Hall, 1985.

7. Martin, James. *Managing the Data Base Environment.* Englewood Cliffs, New Jersey. Prentice-Hall, 1983.

8. Martin, James. *Software Development without Programmers.* Englewood Cliffs, New Jersey. Prentice-Hall, 1984.

9. McFarlan, F. Warren and McKenny, James L. *Corporate Information Systems Management: The Issues Facing Senior Executives.* Homewood, Illinois. Richard D. Irwin, 1983.

10. McLeod, Raymond, Jr. *Decision Support Software for the IBM Personal Computer.* Toronto, Ontario. Science Research Associates, Inc., 1985.

11. Mosevich, J.W. and McCandless, W. *Managing Fourth Generation Technologies: Guidelines for Action. CIPS/ACI Congress 86.* Vancouver, British Columbia. 1986.

12. Myers, Glenford. *The Art of Software Testing.* New York, New York. John Wiley and Sons, 1978.

13. Rockart, John F. and Bullen, Christine V. (editors). *The Rise of Managerial Computing: the Best of the Center for Information Systems Research.* Homewood, Illinois. Dow Jones-Irwin, 1986.

14. Sprague, R.H. and Carlson, E.D. *Building Effective Decision Support Systems.* Englewood Cliffs, New Jersey. Prentice-Hall, 1982.

15. Sprague, Ralph and McNurlin, Barbara. *Information Systems Management in Practice.* Englewood Cliffs, New Jersey. Prentice-Hall, 1986.

16. Weinberg, Gerald M. *Rethinking Systems Analysis and Design.* Boston, Mass. Little, Brown and Company, 1982.

Chapter 11

FOURTH-GENERATION TECHNOLOGIES

11.1 The Five Generations of Software

Underlying many of the problems with the traditional life cycle are its lack of automation and its implied use of computer languages such as COBOL, PL/1 and FORTRAN. There are now emerging new tools which provide some automated support for systems development and which can create applications in a fraction of the time, effort and cost needed with the above languages. We shall now describe the main tools currently in use, and in Chapter 12 we shall show how they are being utilized. We begin with the different types of software presently in use.

11.2 Computer Programming Languages

11.2.1 First-Generation Languages

The first electronic computers were constructed out of vacuum tubes (present-day computers use integrated circuits of microtransistors), they weighed several tons, and consumed vast amounts of electricity. The way in which they were programmed was also quite primitive. At that time (the 1940's) the only way to program a computer was in so-called *machine language*. This was fed into the computer on cards or paper tape, or sometimes wired into large circuit boards. Machine language was, and still is, the only language understood by a computer and if, in the past, an operator or programmer wanted to write out a program on paper to check it or discuss it with someone, it would have been represented as a list of zeros and ones.

Machine language is often referred to as a *first-generation language.* One aspect of machine language is that since the computer can understand and execute it immediately, it is very machine efficient. On the other hand, it is extremely difficult for human beings to understand and create programs. Thus, it is people inefficient, and for this reason, the next generation of computer programming languages was invented.

11.2.2 Second-Generation Languages

The *second generation* of computer programming languages (1950's) uses two-, three- and four-letter codes to represent individual instructions. To illustrate how cumbersome it is, a portion of a program which inputs, adds and outputs two numbers is reproduced below.

```
               ldd      #outprompt  ;display the prompt
               jsr      display ;using display subroutine
               ldd      #pntr   ;get first integer from
               jsr      read    ;the keyboard
               stb      firstnum        ;store the first integer
               ldd      #pntr   ;get the second integer
               jsr      read
               addb     firstnum        ;add firstnum to the second
                                        ;integer stored in b
               clra
               pshs     d       ;store the result on the stack
               ldd      #pntr   ;translate the binary represent
               jsr      itos_   ;ation to a decimal string for
                                ;output
               ldd      #pntr   ;display the result on the
               jsr      display ;screen
               swi
outprompt      fcc      "enter two numbers to add"
pntr           rmb      40
firstnum       fcb      $00
display        tfrd,x   ;initialize the index register
loop           ldb,x+   ;load character byte into b
```

```
beq        quit     ;put characters to the screen until
jsr        putchar_        ;end of string
bra        loop
quit       rts
read       tfrd,x   ;initialize index register
std        index    ;store pointer
loop       jsrgetchar_     ;get each character until
cmpb       #cr      ;carriage return
beq        endloop
stb        ,x+      ;store character in buffer
bra        loop
endloop    ldb#00   ;store end_of_string
           stb,x+
ldd        index    ;convert decimal string to binary
jsr        stoi_    ;representation via library subroutine
rts
index      rmb2
```

Second-generation languages like the one above are called *assembler languages* and are still used to program special operations in both microcomputers and mainframe computers. The example above is the assembler language used on the SuperPet microcomputer. Assembler languages are almost never used for programming business applications because they, like machine language, are human inefficient. However, during the late 1950's, these were the only programming languages available. It is clear how tedious it was to create even simple applications because one had to keep track of memory locations and to specify routine functions such as reading in and printing out data.

Assembler languages are quite machine efficient, but since the computer can only understand machine language, the assembler language program must be translated into machine language. This process is called assembling the program, and is performed by the computer itself.

11.2.3 Third-Generation Languages

During the late 1950's and early 1960's, the development of *third-generation languages* made available much more human-efficient programming capabilities. Examples of third-generation software are BASIC, FORTRAN, and COBOL. There are hundreds of such languages in existence, although only about a dozen are widely known and used. Third-generation languages use English language terms, such as *print*, *read*, *write*, *stop*, as operations. They also use mathematical symbols, such as +, -, *, /, =, <, >, as well as built-in functions, such as AVE, MAX, MIN, LOG, SINE, COSINE and EXP.

The following is a BASIC language version of the above assembler program.

```
INPUT X, Y

LET Z = X + Y

PRINT X, Y, Z

END
```

As you can see, it is a very simple task to create this program as compared to the assembler counterpart.

Each third-generation language is designed to suit some types of applications better than others. For example, FORTRAN (which comes from the phrase **FOR**mula **TRAN**slation) is widely used for engineering and scientific calculations, as it is best suited for numerical computation. FORTRAN is not suited for report writing. COBOL (**CO**mmon **B**usiness **O**riented **L**anguage) is better suited for extracting information and creating reports, and is not appropriate for applications that are calculation intensive.

Many business application programs, such as computing a person's pay and printing a paycheque, are written by professional programmers. Such programs tend to be very long, requiring thousands of lines of code, because the programmer has to instruct the computer on what is desired, as well as having to specify exactly how to go about doing it. For example, to compute a person's pay, the program must read from one or more computer files the name and other identification fields of each employee, the hours worked, the pay rate, the tax rate and other deductions. For the computer to read the appropriate field, the programmer must utilize some access method (e.g. binary search, direct access etc.) and must write a different routine for these standard processes with every new application. A term often used to describe third-generation languages is *procedural* because the programmer must develop the procedures for attaining the desired results (i.e. telling the computer *how* to produce what is required).

11.2.4 Fourth-Generation Languages

During the late 1970's and early 1980's, new nonprocedural languages were developed. These are referred to as *fourth-generation languages*. They are not procedural, which means that one only has to indicate in a program what is desired (and not *how* to do it). For example, to compute and print a person's pay using the fourth-generation language FOCUS, we type simply:

```
TABLE FILE PAYROLL

PRINT PAY AS HOURS * RATE - (TAX + DEDUCTION)

BY EMPLOYEE-NAME
```

The language itself finds the hours, rate, tax and deductions from the file PAYROLL and automatically creates a report or TABLE. In COBOL one would have to lay out what the report would look like, describe exactly the contents of the file and specify procedures to employ when the end of the file had been reached. With fourth-generation languages, however, these things are accomplished automatically for us. Following is a COBOL program which is equivalent to the above FOCUS query. The sequence numbers have been removed to make the program more readable.

```
IDENTIFICATION DIVISION.
PROGRAM-ID. PAYROLL.

ENVIRONMENT DIVISION.
CONFIGURATION SECTION.
SOURCE-COMPUTER. IBM-4341.
OBJECT-COMPUTER. IBM-4341.

INPUT-OUTPUT SECTION.
FILE-CONTROL.
    SELECT EMPLOYEE-PAYROLL-FILE
            ASSIGN TO UT-S-PAYROLL.
    SELECT PAYROLL-REPORT-FILE
            ASSIGN TO UT-S-PRINTER.

DATA DIVISION.
```

```
FILE SECTION.
FD  EMPLOYEE-PAYROLL-FILE
    RECORD CONTAINS 80 CHARACTERS
    LABEL RECORDS ARE STANDARD
    DATA RECORD IS EMPLOYEE-PAYROLL-RECORD.
01  EMPLOYEE-PAYROLL-RECORD.
    04  EMPLOYEE-NAME.
        08  EMPLOYEE-LAST-NAME              PIC X(20).
        08  EMPLOYEE-INITIALS               PIC XXX.
    04  EMPLOYEE-NUMBER                     PIC 9(9).
    04  EMPLOYEE-PAY-DATA.
        08  EMPLOYEE-HOURS-WORKED           PIC 9(3)V9.
        08  EMPLOYEE-PAY-RATE               PIC 9(3)V99.
        08  EMPLOYEE-INCOME-TAX             PIC 9(4)V99.
        08  EMPLOYEE-DEDUCTIONS             PIC 9(4)V99.
    04  FILLER                              PIC X(27).

FD  PAYROLL-REPORT-FILE
    RECORD CONTAINS 133 CHARACTERS
    LABEL RECORD IS OMITTED
    DATA RECORD IS PAYROLL-REPORT-RECORD.
01  PAYROLL-REPORT-RECORD.
    04  PAY-REPORT-CARRIAGE-CONTROL         PIC X.
    04  FILLER                              PIC X(4).
    04  PAY-REPORT-NAME.
        08  PAY-REPORT-LAST-NAME            PIC X(20).
        08  PAY-REPORT-INITIALS             PIC XXX.
    04  FILLER                              PIC X(4).
    04  PAY-REPORT-NET-PAY                  PIC ZZ,ZZZ.99.
    04  FILLER                              PIC X(92).

WORKING-STORAGE SECTION.
01  PROGRAM-INDICATORS.
    04  END-OF-FILE-INDICATOR               PIC XXX VALUE 'NO '.
    04  NUMBER-OF-LINES-PRINTED             PIC 99  VALUE 0.

01  PAY-CALCULATIONS.
    04  GROSS-PAY                           PIC 9(5)V99.
    04  INCOME-TAX                          PIC 9(4)V99.
    04  GROSS-PAY-LESS-TAX                  PIC 9(5)V99.
    04  DEDUCTIONS                          PIC 9(4)V99.
    04  NET-PAY                             PIC 9(5)V99.
```

```
01 REPORT-HEADING-LINE.
    04 REPORT-HEADING-CARRIAGE-CTRL    PIC X.
    04 FILLER                          PIC X(4) VALUE SPACES.
    04 FILLER                          PIC X(13)
        VALUE 'EMPLOYEE NAME'.
    04 FILLER                          PIC X(14) VALUE SPACES.
    04 FILLER                          PIC X(7)
        VALUE 'NET PAY'.
    04 FILLER                          PIC X(94) VALUE SPACES.

PROCEDURE DIVISION.

OPEN-FILES.
    OPEN INPUT EMPLOYEE-PAYROLL-FILE
        OUTPUT PAYROLL-REPORT-FILE.

    READ EMPLOYEE-PAYROLL-FILE AT END
        MOVE 'YES' TO END-OF-FILE-INDICATOR.

PRODUCE-REPORT.
    PERFORM CALCULATE-PAY THRU READ-NEXT-RECORD
        UNTIL END-OF-FILE-INDICATOR EQUAL 'YES'.

CLOSE-FILES-AND-STOP.
    CLOSE EMPLOYEE-PAYROLL-FILE
          PAYROLL-REPORT-FILE.
    STOP RUN.

CALCULATE-PAY.
    MULTIPLY EMPLOYEE-HOURS-WORKED BY EMPLOYEE-PAY-RATE
        GIVING GROSS-PAY.

    MOVE EMPLOYEE-INCOME-TAX TO INCOME-TAX.
    MOVE EMPLOYEE-DEDUCTIONS TO DEDUCTIONS.

    SUBTRACT INCOME-TAX FROM GROSS-PAY
        GIVING GROSS-PAY-LESS-TAX.

    SUBTRACT DEDUCTIONS FROM GROSS-PAY-LESS-TAX
        GIVING NET-PAY.
```

```
WRITE-REPORT-RECORD.
    PERFORM HEADING-CHECK.

    MOVE SPACES TO PAYROLL-REPORT-RECORD.
    MOVE EMPLOYEE-NAME TO PAY-REPORT-NAME.
    MOVE NET-PAY TO PAY-REPORT-NET-PAY.
    WRITE PAYROLL-REPORT-RECORD.

    ADD 1 TO NUMBER-OF-LINES-PRINTED.

READ-NEXT-RECORD.
    READ EMPLOYEE-PAYROLL-FILE AT END
        MOVE 'YES' TO END-OF-FILE-INDICATOR.

HEADING-CHECK.
    IF  NUMBER-OF-LINES-PRINTED EQUAL ZERO
     OR NUMBER-OF-LINES-PRINTED GREATER THAN 60
          MOVE '1' TO REPORT-HEADING-CARRIAGE-CTRL
          MOVE REPORT-HEADING-LINE TO PAYROLL-REPORT-RECORD
          WRITE PAYROLL-REPORT-RECORD
          MOVE SPACES TO PAYROLL-REPORT-RECORD
          WRITE PAYROLL-REPORT-RECORD
          MOVE 2 TO NUMBER-OF-LINES-PRINTED.
```

Note that not only is the FOCUS query much shorter than the COBOL program, but it is also very easy to understand. The COBOL program would require much study in order to comprehend its function. It can be seen that programs written in fourth-generation languages are more human efficient and therefore easier to develop and maintain.

Fourth-generation languages are fairly new but are being used more and more frequently in many businesses. They appear to offer substantial benefits over third-generation languages because of how quickly a person can write a program. In fact, people who are not programmers can use them. Just as we saw with third-generation languages, there are differences among fourth-generation languages, making some better than others for certain applications.

One of the most successful classes of business software packages is the so-called electronic spreadsheet. Basically, all one does is enter numbers into *cells* which are locations on the electronic sheet, specify the arithmetic operations and then just let the computer do, for example, a budget. Examples of popular spreadsheets are Lotus 1-2-3, MultiPlan, VP Planner, and SuperCalc.

11.2.5 The Fifth Generation

The last generation of computer software is, obviously, the *fifth generation*, although strictly speaking there are no fifth-generation languages as there are fourth-generation. With all of the previous generations, we must follow strict rules when creating programs. In the fifth-generation the computer can interpret our own language (e.g., English, French, Spanish, etc.) and produce a desired result. For example, suppose we needed a list of all male and female students in our school who are over ten years of age. In a fifth-generation language, we just wrote the program. All we need to type is:

> "List all male and female students in our school who are over 10 years of age"

The computer would have the ability to comprehend that "our school" is the one attended by the person asking. It would then search its file, or files, for males and females over ten and print the results. In a third- or fourth-generation language, the computer would probably indicate that no one is male and female. In other words, it would not have the *intelligence* to understand what we really meant.

Fifth-generation computing is concerned with such areas as artificial intelligence, expert systems, robotics and voice recognition. Today, with the exception of robotics, these tecnologies have so far accomplished little penetration into the business world, but within the next decade, judging by their potential, they should be valuable tools in many areas.

11.2.6 Special Purpose Languages

There are some third- and fourth-generation computer languages that were created for very specialized applications. These languages, called special purpose languages, fit somewhere between general purpose languages (FORTRAN, COBOL, FOCUS, etc.) and the software packages discussed in the next section. Special purpose languages offer an alternative to writing a whole new application or purchasing a packaged one that may be too sophisticated or expensive.

The distinguishing feature of a special purpose language is that, as the name implies, it offers features devoted to specific functions, such as statistical computations, forecasting, report writing, word and text processing, mathematical computations and simulation. The main benefit, and perhaps the reason for the existence of special purpose languages, is that in almost all business, scientific and engineering applications, many functions are the same. For example, the problem of computing averages (marks, pay, height, weight, age, etc.) requires using the same basic formula. So there is no need to reprogram the computation. It is more convenient to type simply:

```
Y = AVG (PAY)
```

or a similar statement. In addition to the convenience of special purpose languages, we can be confident of accuracy because such software is thoroughly tested by the thousands of companies that use it. Some examples of special purpose languages follow.

One of the first areas to benefit from special purpose languages was statistics. Many companies compute their own and their competitors' averages and trends and consumer trends. SAS is a fourth-generation language devoted mainly to statistical and forecasting applications. It offers many easy-to-use features such as report inventory and graphics. For example, to compute the average (mean) marks of all students in a class, we could use the SAS program shown in Figure 97.

```
DATA GRADES;
INPUT STUDENT$ MARK;
CARDS;
      J. SMITH   35   (data)
      M. JONES   37
          .
          .
          .
PROC MEANS; (compute averages)
PROC PRINT; (print results)
```

Figure 97: SAS program

As you can see, SAS is very powerful and much easier than having to write an original program.

Another area using special purpose languages is *simulation*. Simulation means having a program mimic what occurs in reality by using mathematical formulae and random number generators. For example, if a supermarket manager needs to decide how many checkout positions to staff during each hour of the day, it is possible to simulate the daily pattern of customer arrivals. This can be done very simply with a simulation language such as SIMSCRIPT or GPSS (General Purpose System Simulator). Before the development of these languages, we would have had to write a new FORTRAN program for each simulation problem. Someone noticed that the vast majority of simulation problems use very similar functions, and from this observation grew many simulation languages. A simple example in GPSS follows in which the arrival of customers to a supermarket is simulated.

```
GENERATE  36,FN1       CREATE SHOPPERS
TRANSFER  BOTH,,AWAY   CHECK FOR AVAILABLE BASKET
ENTER     BSKT         GET A BASKET
ASSIGN    1,FN2        DETERMINE NO. OF ITEMS
ADVANCE   1,FN3        SHOP
QUEUE     WAIT         WAIT FOR COUNTER SPACE
ENTER     CKT          GET COUNTER SPACE
DEPART    WAIT         LEAVE QUEUE
ADVANCE   V1           CHECK-OUT
LEAVE     CKT          FREE COUNTER SPACE
TABULATE  TRT          TABULATE TRANSIT TIME
TABULATE  ITM          TABULATE NO. OF ITEMS
LEAVE     BSKT         RETURN BASKET
TERMINATE              1
```

The output is a report showing the average length of lineups and the length of time a customer stands in them.

These are examples from only two of the many areas that commonly use special purpose software. Keep in mind that with any language, no matter what its generation or purpose, the business person must have a program written, or, if possible, must write a new program. With special purpose or fourth-generation languages it is vastly easier to write a program because what are normally complicated procedures (e.g., file access) or computations (e.g., standard deviation) are single commands. Even this can be time consuming, but sometimes it is the only possibility. There is, however, a third class of software which requires no programming at all.

11.2.7 Packaged Software

You can now walk into any computer store and purchase a diskette containing a complete, ready-to-use application. All you have to do is follow instructions. One example you are already familiar with is the computer game. A game is, after all, a computer program. But the player did not have to write it; it is only necessary to provide the input using joysticks and buttons. There are now available many business application packages referred to as packaged software or off-the-shelf software.

Business people can purchase ready-to-use word processing, accounting, payroll, membership lists, personal filing and inventory packages; the list goes on and on. All you have to do is read any popular magazine on personal computing to see just how many such packages are available. In fact, the current trend in organizations is increasingly toward the use of packaged software, not merely for personal computers, but for minis and mainframes, and for applications at all levels of sophistication. Packages save a great deal of time as you do not have to wait for someone to write a program. They do cost money, but so do programmers. The main fault of packaged software is that sometimes it does not do things exactly the way one would like. But as time goes on, more and more flexible packages (called parametrized packages) are becoming available that can be tailored or altered for a particular situation.

11.2.8 Choosing The Best Option

Having had an overview of the many options available to the business person, you may wonder how the best option is selected. That will depend on the type of application and on the person who will be using the computer. A rough guideline follows:

1. If the application is fairly standard, then packaged software is probably available. It must naturally be compatible (able to be run) with the computer currently in use by the business. There will be a cost to purchase the package and some study or training may be required. One possible problem in using some packages is that sometimes they are not compatible among themselves. For example, a word processing package may not be capable of incorporating a spreadsheet file so that it can be included in a report. To do this it is necessary to purchase compatible packages.

2. If a package is not appropriate because one's requirements change often, or are fairly special, then a special purpose language, or fourth-generation language, will probably offer a solution. If the special purpose language is fourth-generation, then little programming may be required, and the business people may be able to create their own applications. Other times, a programmer can usually create the application program much faster and with fewer errors than with FORTRAN or COBOL by using the special purpose or fourth-generation language.

3. If an application is unique or requires especially fast response time, the only option may be to have it programmed in a third- or even second-generation language. This is appropriate for reservation systems for airlines, hotels or car rental agencies; automatic teller banking systems; and any so-called on-line transaction system. These applications are very complex: they involve the immediate updating of computer files and they must be precisely defined. Fourth-generation and special purpose software are currently too slow or not flexible enough for such applications.

In summary, a good approach is: if you can do so, purchase; if not, investigate special purpose or fourth-generation languages; and, as a last resort, program.

11.3 Fourth-Generation Languages

Fourth-generation languages have been available for about ten years but only recently have they come into widespread use. As we shall now see, they offer substantial productivity gains to data processing personnel. They also offer the potential for end-user computing, a current use in many corporations. In addition to these advantages, fourth-generation languages have a great potential in systems analysis: by their ability to quickly create simple applications or test versions of complicated applications. Their uses will be discussed in Chapter 12.

There seems to be no technical definition of fourth-generation language, but a set of attributes is often listed by various authors to describe them. We feel that the set given by James Martin [3] suits our needs best. It is reprinted here with permission.

- It is user friendly.

- A nonprofessional programmer can obtain results with it.

- It employs a database management system directly.
- Procedural code requires instructions that are an order of magnitude fewer than COBOL.
- Nonprocedural code is used where possible.
- It makes intelligent default assumptions about what the user wants, where possible.
- It is designed for on-line applications.
- It enforces or encourages structured code.
- It is easy to understand and maintain another person's code.
- Non-DP users can learn a subset of the language in a two-day training course.
- It is designed for easy debugging.
- Results can be obtained in substantially less time than with COBOL or PL/1.

In the computer trade press, many commercially available fourth-generation products are refered to as *application generators* and some as *query languages* depending on their power and sophistication. We feel that it is unnecessary to make such distinctions in this text and so will refer to all such products as fourth-generation languages or 4GL's. Following is a partial list of several 4GL's and DBMS's currently in wide use:

Product	Vendor
ADS/O	Cullinet Software, Inc.
DB2, SQL/DS	IBM Corporation
FOCUS	Information Builders, Inc.
Mapper	Sperry Corporation
MANTIS, SUPRA, ULTRA	Cincom Systems
Natural	Software AG of N.A.
NOMAD2	Must Software International
ORACLE	Oracle Corporation
POWERHOUSE	Cognos Inc.

Ramis	On-Line Software International, Inc.
SAS	SAS Institute Inc.
TELON	Pansophic Systems, Inc.

Each of these products has its own strengths and weaknesses. They often incorporate their own DBMS's, data dictionaries and features such as graphics and data communication for personal computer links. Also, they usually have interfaces with other systems or DBMS's such as IMS.

Fourth-generation languages are being implemented in more and more companies and they are expected to have a substantial impact on systems development and in particular on systems analysis.

11.3.1 Fourth-Generation Language Performance

Before we proceed with our detailed introduction to a specific 4GL we feel it necessary to mention the issue of performance and computer resource consumption of 4GL's. When one first broaches the subject of 4GL's to a systems professional, the first concern mentioned will undoubtedly be 4GL impact on hardware resources. Fourth-generation languages, being powerful interpretive languages, require more concurrent CPU resources than compiled languages such as COBOL. Further, we can expect that increasing the exposure of the system to many end-users in effect creates new users, thus increasing demand for CPU, I/O channels and disk space.

These are certainly valid concerns, and must not be taken lightly. However, the computer exists to serve the business, and if more resources are needed, so be it, as long as they can be justified from the viewpoint of benefits (justified by the users, of course). All 4GL's are continually being improved by their creators, and performance usually improves with each subsequent release. Also, performance is often a function of query or application structure and data structure. This means that once an application is running and giving satisfactory results it can often be improved by redesign and analysis. Finally, any fair comparison between third- and fourth-generation languages should take into account the time required for compiling and debugging and using CPU resources, and most of all human resource and elapsed-time considerations. These tend to put 4GL's in a very good light.

In conclusion, there is no fundamental reason why the performance of 4GL's cannot compare with 3GL's if they are managed and used intelligently. Their benefits should be defined not just from the standpoint of resource consumption, but also from the corporate or business viewpoint.

11.4 An Introduction To A Specific Fourth-Generation Language: NOMAD2

In order to provide insight into the capabilities of a typical 4GL we shall now present an introduction to NOMAD2. This is a fourth-generation language and relational database management system powerful enough to be classified as an application generator. It is a nonprocedural language but it does have a procedural subset, and it provides relational as well as hierarchical logical views of data. In addition, it has the features present in other 4GL's such as graphics and an IMS interface as well as a decision support feature for goal seeking and what-if analysis. Note that our presenting NOMAD2 is not to be interpreted as an endorsement. Each product has its strengths and weaknesses and should be evaluated accordingly. For example, like NOMAD2, FOCUS and RAMIS II are based on hierarchical database design and they too provide relational operations and views. They also have graphics and all three provide user-friendly interfaces for novice users.

Perhaps the main characteristics of 4GL's are the following:

- the data is defined separately from the procedures that use it, and therefore, stands alone
- the programs (procedures) are nonprocedural and written in language approximating English

These points should be evident in the examples and explanations which follow.

Fourth-generation languages can accelerate the development effort by an order of magnitude when compared to development using third-generation technologies. The starting point in software development using fourth-generation technologies is the analysis of the user's data requirement.

NOMAD2's DBMS is based on the relational model, but as will be discovered, NOMAD2 also has the flexibility to accommodate hierarchical views.

11.4.1 Data Definition - Creating the Schema

Recall from Chapters 3 and 4 the notions of conceptual data modeling using entity-relationship diagrams, and their translation into the relational model. Figure 14 in Chapter 3 depicted a conceptual model of a portion of a university and is repeated in Figure 98. Using NOMAD2, the database schema can be derived almost directly from the E-R model, and even more directly using the relational equivalent.

Suppose for this example that one initially wishes to store information about the entities FACULTIES, DEPARTMENTS, STUDENTS, PROFESSORS and COURSES. Figure 99 shows a partial listing of possible attributes corresponding to each entity set. Translation into NOMAD2 follows almost directly. A NOMAD2 schema corresponding to Figure 99 is shown in Figure 100.

A schema begins with the word SCHEMA, and, as is evident, each entity (or relation) is referred to as a *master* in NOMAD2.

The keys of each relation are indicated by "keyed (itemname)":

- the key for FACULTIES is Facname, the name of the faculty
- the key for DEPARTMENTS is Deptname, department name
- the key for STUDENTS is Studid, student identification number
- the key for PROFESSORS is Profid, professor's employee number
- the key for COURSES is the composite key Coursename Courseno, course department plus course number (e.g., English 102).

The schema terminates with the line

PERSPECT;

The PERSPECT statement allows for the entry of data into NOMAD2 through a screen-oriented database editor, DBEDIT, which provides interactive access to a database. Think of the time this saves compared to traditional ways of entering data into a set of files! With the traditional approach, one must first

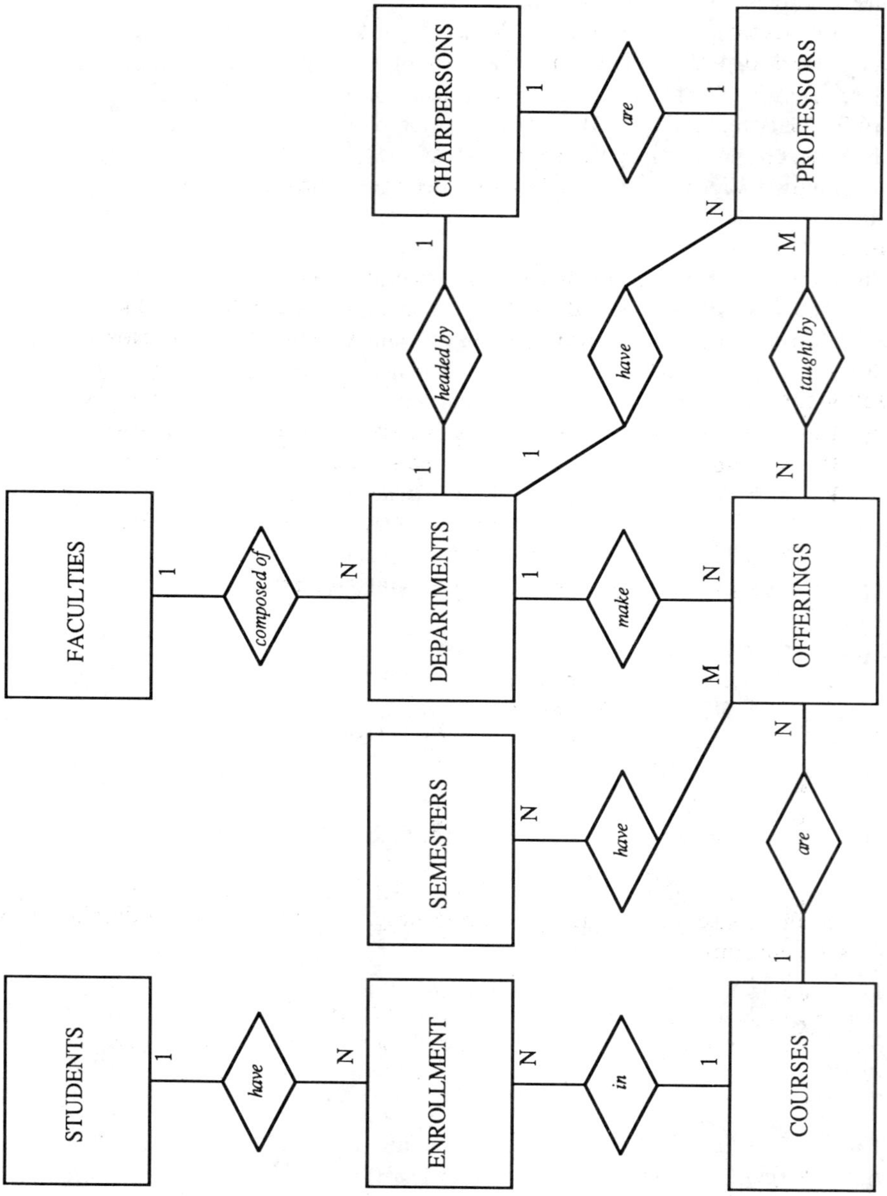

Figure 98: Partial conceptual model of a university

```
FACULTIES
      Facname     - faculty name
      Facbudget   - annual budget

DEPARTMENTS
      Deptname    - department name
      Deptbudget  - annual department budget

STUDENTS
      Studid      - student's identification number
      Studname    - student's last name and initials
      Studmajor   - major program of study (department)

PROFESSORS
      Profid      - professor's identification number
      Profname    - professor's last name and initials
      Profrank    - professor's rank
                    (full, associate, visiting, etc.)
      Profsalary  - professor's annual salary
      Profdept    - professor's department

COURSES
      Coursename  - course department
      Courseno    - course number
      Coursedesc  - course description
```

Figure 99: Partial listing of attributes

define the database within the application program and then write detailed procedures for accepting the data.

The schema is defined in a file using a text editor, and then must be compiled. This process provides for the framework within which data can be entered and reports written.

11.4.2 Reporting

Display formats for each item (field) are given within the schema, for example,

- Studid is defined to be four integers in length

```
SCHEMA;

master FACULTIES keyed(Facname);
        item Facname a20;
        item Facbudget as $999,999,999;

master DEPARTMENTS keyed(Deptname);
        item Deptname a12;
        item Deptbudget as $9,999,999;
        item Deptfac member'FACULTIES';

master STUDENTS keyed(Studid);
        item Studid 9999;
        item Studname a20;
        item Studmajor member'DEPARTMENTS';

master PROFESSORS keyed(Profid);
        item Profid as format'999-999'
             heading'EMPLOYEE:NUMBER';
        item Profname a20;
        item Profrank a5 heading'PROF:RANK';
        item Profsalary as $99,999 heading'SALARY';
        item Profdept member'DEPARTMENTS';

master COURSES keyed(Coursename,Courseno);
        item Coursename member'DEPARTMENTS';
        item Courseno 999 heading'NO';
        item Coursedesc a30;

PERSPECT;
```

Figure 100: NOMAD2 schema

- Studname is twenty characters
- the budget items Facbudget and Deptbudget are displayed with currency formatting
- the item Profid is six integers long, separated by a hyphen into two groups of three.

When reports are printed using this schema, headings will, by default, use the item name as the heading. For example, the NOMAD2 command

LIST BY DEPTFAC BY DEPTNAME DEPTBUDGET

produces the report shown in Figure 101.

```
                                                   PAGE    1

DEPTFAC                     DEPTNAME          DEPTBUDGET
--------------------        ------------      ----------
ARTS                        BUSINESS          $4,200,000
                            ENGLISH           $2,690,000
                            FRENCH            $1,450,000
ENGINEERING                 ELEC ENGRING      $3,005,000
                            MECH ENGRING      $2,742,000
SCIENCE                     COMPUTER SCI      $3,210,000
                            MATHEMATICS       $4,400,000
```

Figure 101: LIST BY DEPTFAC BY DEPTNAME DEPTBUDGET

The BY option of the LIST command and the item following cause the report to be ordered according to that item (called a BY-item). Notice that in a report, duplicate items are not listed. Each faculty has several departments, but each faculty is named only once in the report.

The item Profid in the master PROFESSORS will have the column heading EMPLOYEE NUMBER because of the heading option specified within the schema. The contents of a master may be obtained by the command LIST mastername, as for example, LIST PROFESSORS (Figure 102).

Notice the item entries within the schema for the fields for Deptfac, Studmajor, Profdept and Coursename:

```
item Deptfac member'FACULTIES';
item Studmajor member'DEPARTMENTS';
item Profdept member'DEPARTMENTS';
item Coursename member'DEPARTMENTS';
```

PAGE 1

EMPLOYEE NUMBER	PROFNAME	PROF RANK	SALARY	PROFDEPT
112-334	HILL, V.J.	FULL	$60,000	COMPUTER SCI
123-456	MORANSKI, J.W.	FULL	$58,000	MATHEMATICS
146-892	SAMPSON, R.B.	ASSOC	$55,000	MATHEMATICS
238-945	PRIDE, J.G.	ASST	$50,600	ENGLISH
390-800	BROWN, J.F.	ASSOC	$52,400	FRENCH
405-192	HUTT, S.E.	FULL	$61,000	BUSINESS
654-448	ROBINSON, N.	ASSOC	$55,000	COMPUTER SCI
667-890	TROUSSLER, M.B.	ASSOC	$58,000	BUSINESS
678-631	CRAIG, H.G.	ASST	$53,400	ELEC ENGRING
753-994	DREW, S.C.	ASSOC	$54,100	ELEC ENGRING
841-325	GREER, G.E.	ASSOC	$48,800	FRENCH
879-954	WEXLER, R.L.	FULL	$59,000	ENGLISH
977-320	JOHNSTON, J.G.	ASSOC	$52,600	MATHEMATICS
990-765	CASSON, L.C.	ASSOC	$54,300	FRENCH

Figure 102: LIST PROFESSORS

These descriptions mean that any value of Deptfac must be an existing value found in the entity FACULTIES and that any instances of Studmajor, Coursename and Profdept must correspond to an instance in the entity DEPARTMENTS. Such a powerful feature is called *referential integrity* and relieves us of the task of writing applications programs that perform integrity checks. (The fields Deptfac, Studmajor, Profdept and Coursename are connection fields to the fields Facname and Deptname respectively. Their names have been altered in order that they be unique. It will therefore not be necessary to qualify these attributes according to the particular master being reported.)

NOMAD2 can produce totals and subtotals automatically.

LIST BY PROFDEPT PROFNAME PROFSALARY SUBTOTAL TOTAL

produces the report shown in Figure 103.

```
                                          PAGE    1

PROFDEPT        PROFNAME               PROFSALARY
------------    --------------------   ----------
BUSINESS        HUTT, S.E.                $61,000

                TROUSSLER, M.B.           $58,000
                                       ----------
*                                        $119,000

COMPUTER SCI    HILL, V.J.                $60,000

                ROBINSON, N.              $55,000
                                       ----------
*                                        $115,000

ELEC ENGRING    CRAIG, H.G.               $53,400

                DREW, S.C.                $54,100
                                       ----------
*                                        $107,500

ENGLISH         PRIDE, J.G.               $50,600

                WEXLER, R.L.              $59,000
                                       ----------
*                                        $109,600

FRENCH          BROWN, J.F.               $52,400

                CASSON, L.C.              $54,300

                GREER, G.E.               $48,800
                                       ----------
*                                        $155,500

MATHEMATICS     JOHNSTON, J.G.            $52,600

                MORANSKI, J.W.            $58,000

                SAMPSON, R.B.             $55,000
                                       ----------
                                         $165,600
                                       ==========
                                         $772,200
```

Figure 103: Totals in NOMAD2

11.4.3 Entering Data after the Schema has been Defined

Suppose we have created the schema described above, and are now ready to enter data using the screen-oriented database editor, DBEDIT. Invoking DBEDIT causes the screen shown in Figure 104 to appear.

```
Please select what you would like to edit ##

   1. FACULTIES
   2. DEPARTMENTS
   3. STUDENTS
   4. PROFESSORS
   5. COURSES
```

Figure 104: DBEDIT screen

To enter data into a master, for example, STUDENTS, response number 3 is given, and a screen appears with the names of each field at the top (see Figure 105).

One can enter as much data and cursor around the screen as much as required. The editor performs data screening and integrity checking as described earlier. *Data screening* refers to the act of defining what characterizes valid data. Must it be numeric or alphabetic, or does it have a defined pattern, for example, Universal Product Code (UPC)? Are there limits on the range of values that may be associated with a data item? All of these characteristics may be specified in the definition of a database, and used to automatically verify the accuracy of the data to be entered. For example, if a nonnumeric character is entered in the item Studid, or if a nonexistent department is

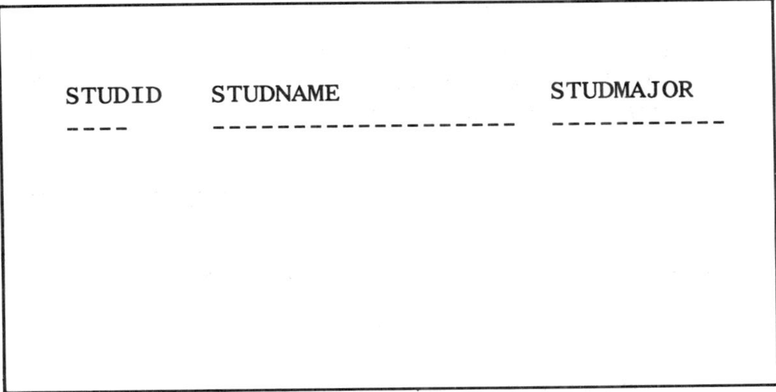

Figure 105: STUDENTS screen in DBEDIT

entered into the item Studmajor, an error message appears on the screen and the cursor is positioned at the offending value.

11.4.4 Refining the Database Structure

Assume the required data have been entered, and that various reports have been obtained. Although the information is useful, it would be more useful if it contained information relating to semesters, course offerings and instructors, and student marks. With a third-generation language, such a refinement of the database would require considerable effort. However, in NOMAD2, to add such information one needs only to enter the text editor and edit the schema, adding a master called SEMESTERS, and information regarding course offerings, instructors and student marks. After the modifications have been completed, the schema must be recompiled. This process of refining a database structure is similar to the process of modifying a computer program. Another aspect which makes the process of database refinement a relatively easy task is that the existing data remain intact, and are not destroyed by the structural reorganization. The revised schema is shown in Figure 106.

There are several aspects of this revised schema worth mentioning. Notice that the masters FACULTIES, DEPARTMENTS, STUDENTS and PROFESSORS remain the same. The master SEMESTERS has been inserted into a convenient position within the schema. It has only one attribute, Semes-

```
SCHEMA;

master FACULTIES keyed(Facname);
       item Facname a20;
       item Facbudget as $999,999,999;

master DEPARTMENTS keyed(Deptname);
       item Deptname a12;
       item Deptbudget as $99,999,999;
       item Deptfac member'FACULTIES';

master SEMESTERS keyed(Semester);
       item Semester a3 heading'SEM';

master STUDENTS keyed(Studid);
       item Studid 9999;
       item Studname a20;
       item Studmajor member'DEPARTMENTS';

master PROFESSORS keyed(Profid);
       item Profid as format'999-999'
            heading'EMPLOYEE:NUMBER';
       item Profname a20;
       item Profrank a5 heading'PROF:RANK';
       item Profsalary as $99,999 heading'SALARY';
       item Profdept member'DEPARTMENTS';

master COURSES keyed(Coursename,Courseno);
       item Coursename member'DEPARTMENTS';
       item Courseno 999 heading'NUM';
       item Coursedesc a30;

segment OFFERINGS parent COURSES keyed(Semesternum);
       item Semesternum member'SEMESTERS' heading'SEM';
       item Instructor member'PROFESSORS';
       define Instname extract'Profname from PROFESSORS key
              Instructor';

       segment CLASSLIST parent OFFERINGS;
              item Student member'STUDENTS' heading'STUD';
              define Stname extract'Studname from STUDENTS
                     key Student';
              item Grade 999 limits=(0:100);
PERSPECT;
```

Figure 106: Revised university schema

ter. Examples of values might be F87, W88, S88. (A complete listing of the data for each master appears at the end of this chapter in Figures 120 to 125.)

The revised schema illustrates the ability of NOMAD2 to provide a hierarchical as well as a relational view of data. Courses may or may not be offered during any semester; in fact, OFFERINGS may be viewed as "weak" or existence dependent upon COURSES. For this reason, the authors have chosen to introduce the SEGMENT statement. Its purpose is to define a subgrouping within a master. The revised schema has two such subgroupings:

- one for course OFFERINGS within COURSES (keyed by Semesternum),
- the other for CLASSLIST within OFFERINGS (not keyed).

```
segment OFFERINGS parent courses keyed(Semesternum);
        item Semesternum member'SEMESTERS';
        item Instructor member'PROFESSORS';
        define Instname extract'Profname from PROFESSORS key
                Instructor';

        segment CLASSLIST parent OFFERINGS;
                item Student member'STUDENTS';
                define Stname extract'Studname from STUDENTS
                        key Student';
                item grade 999 limits=(0:100);
```

COURSES, OFFERINGS and CLASSLIST are illustrated conceptually in Figure 107.

Although one of the objectives in database design is to minimize data redundancy, there may be occasions when it is convenient to have such redundant data. In this example, we want to have the instructor's name (Instname) to be part of the segment OFFERINGS, even though it "belongs" to the master PROFESSORS, and to have the student's name as part of the segment CLASSLIST, even though it is part of the master STUDENTS. The DEFINE statement provides the capability of defining names which are not actually part of a master or segment, but can be derived from other masters if requested on a report [5]. The items of information are not actually stored with each segment, but are retrieved from the appropriate masters.

Finally, a grade, to be valid, must be in the range 0-100; these limits have been included in the description of the item Grade.

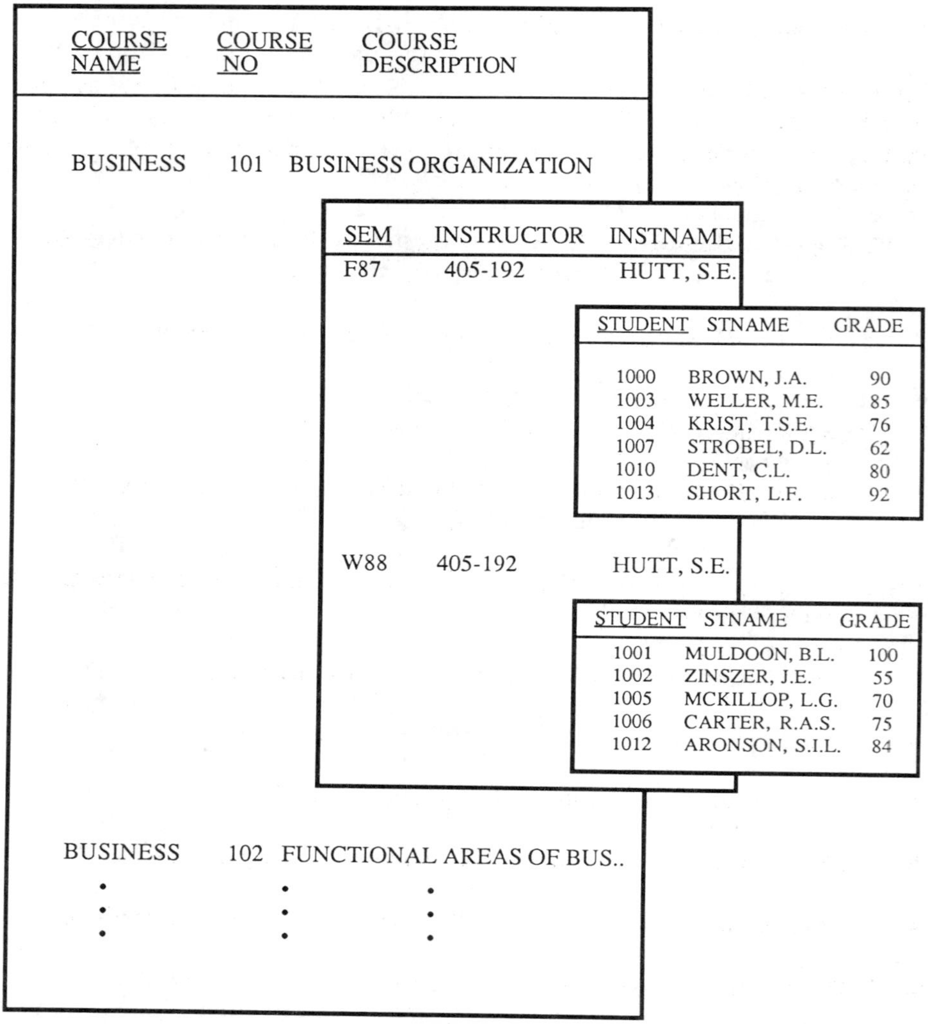

COURSE NAME	COURSE NO	COURSE DESCRIPTION
BUSINESS	101	BUSINESS ORGANIZATION

SEM	INSTRUCTOR	INSTNAME
F87	405-192	HUTT, S.E.

STUDENT	STNAME	GRADE
1000	BROWN, J.A.	90
1003	WELLER, M.E.	85
1004	KRIST, T.S.E.	76
1007	STROBEL, D.L.	62
1010	DENT, C.L.	80
1013	SHORT, L.F.	92

SEM	INSTRUCTOR	INSTNAME
W88	405-192	HUTT, S.E.

STUDENT	STNAME	GRADE
1001	MULDOON, B.L.	100
1002	ZINSZER, J.E.	55
1005	MCKILLOP, L.G.	70
1006	CARTER, R.A.S.	75
1012	ARONSON, S.I.L.	84

COURSE NAME	COURSE NO	COURSE DESCRIPTION
BUSINESS	102	FUNCTIONAL AREAS OF BUS..
•	•	•
•	•	•
•	•	•

Figure 107: COURSES - OFFERINGS - CLASSLIST

11.4.5 Built-in Reporting Functions

Suppose the schema has been recompiled with the above changes and that data have been entered into SEMESTERS, OFFERINGS and CLASSLIST. Information for this database can be reported and summarized using the many reporting functions of NOMAD2. For example,

```
LIST BY SEMESTERNUM BY COURSENAME BY COURSENO BY
                    STNAME GRADE
```

produces a list of grades for every course (Figures 108 and 109).

To obtain the class average for each course,

```
LIST BY SEMESTERNUM BY COURSENAME BY COURSENO
              INSTNAME AVG(GRADE)
```

produces the report shown in Figure 110. Other commonly-used reporting functions include those listed below.

- COUNT counts the number of instances of the item named.
- FIRST retrieves the first instance of the item specified.
- LAST retrieves the last instance of the given item.
- MAX returns the maximum of the values of the item named.
- MEDIAN computes the median of the values for the particular item.
- MIN returns the minimum of the values of the item indicated.
- STDDEV computes the standard deviation of the values of the item given.
- SUM totals the values of the particular item.
- UNIQUE returns the number of distinct values of the given item.
- VARIANCE calculates the variance of the values of the item indicated.

PAGE 1

SEM	COURSENAME	NUM	STNAME	GRADE
F87	BUSINESS	101	BROWN, J.A.	90
			DENT, C.L.	80
			KRIST, T.S.E.	76
			SHORT, L.F.	92
			STROBEL, D.L.	62
			WELLER, M.E.	85
		102	BROWN, J.A.	88
			CARTER, R.A.S.	69
			DENT, C.L.	81
			WELLER, M.E.	75
			ZINSZER, J.E.	73
	COMPUTER SCI	101	ARONSON, S.I.L.	60
			BROWN, J.A.	55
			DENT, C.L.	76
			KRIST, T.S.E.	72
			STROBEL, D.L.	88
	ENGLISH	101	BROWN, J.A.	65
			CARTER, R.A.S.	60
			KRIST, T.S.E.	56
			MCKILLOP, L.G.	98
			MULDOON, B.L.	71
			STROBEL, D.L.	68
			WELLER, M.E.	78
			ZINSZER, J.E.	80
	MATHEMATICS	101	BROWN, J.A.	60
			CARTER, R.A.S.	76
			DENT, C.L.	73
			KRIST, T.S.E.	56
			MCKILLOP, L.G.	81
			WELLER, M.E.	22
F88	BUSINESS	102	MORANSKI, R.	77
			MULDOON, B.L.	56
			RAUCHWAY, A.E.	80
			STROBEL, D.L.	60
	COMPUTER SCI	102	ARONSON, S.I.L.	98
			CARTER, R.A.S.	90
			KRIST, T.S.E.	84

```
MULDOON, B.L.                91
RAUCHWAY, A.E.               73
ZINSZER, J.E.               100
```

Figure 108: LIST BY SEMESTERNUM BY COURSENAME - part 1

Items are sequenced vertically using the BY statement, and can be listed horizontally as well using the ACROSS statement. The NOMAD2 command

LIST BY PROFDEPT ACROSS PROFRANK AVG(PROFSALARY)

produces a matrix-like report, with vertical sequencing according to Profdept and horizontal sequencing according to Profrank (Figure 111). N/A means not available, and is displayed because no data have been entered for some department/rank combinations.

11.4.6 Relational Algebra in NOMAD2

Many more reporting capabilities are available than can be described here. However, several important features warrant discussion. Recall the discussion of relational algebra in Chapter 4. Relational algebra defines the operations permitted on relations. Some of the relational algebra operators of which NOMAD2 is capable include

- *select* - selecting certain rows of relations based on some criteria, for example, for student 1007. This may be done in NOMAD2 in several ways. One way is within the LIST command,

  LIST BY SEMESTERNUM BY COURSENAME BY COURSENO GRADE -
  WHERE STUDENT = 1007 -
  TITLE 'GRADE REPORT FOR STUDENT ' STNAME '   ID:' STUDENT

  This command produces a grade report for the student whose student number is 1007, sequenced by course within semester. The dash at the end of a command line indicates continuation. The report is shown in Figure 112.

- *project* - extracts certain attributes of relations. This is done simply by specifying the items to be included within a report (see Figure 113).

  LIST PROFNAME PROFDEPT

SEM	COURSENAME	NUM	STNAME	GRADE
F88	MATHEMATICS	101	MORANSKI, R.	66
			MULDOON, B.L.	45
			STROBEL, D.L.	80
			ZINSZER, J.E.	87
S88	BUSINESS	102	ARONSON, S.I.L.	84
			KRIST, T.S.E.	79
			SHORT, L.F.	65
	COMPUTER SCI	101	ARONSON, S.I.L.	72
			KRIST, T.S.E.	68
			SHORT, L.F.	70
		102	BROWN, J.A.	69
			STROBEL, D.L.	80
			WELLER, M.E.	77
	ENGLISH	101	ARONSON, S.I.L.	63
			BROOKS, F.	86
			DENT, C.L.	89
			MORANSKI, R.	70
			RAUCHWAY, A.E.	91
			SHORT, L.F.	35
W88	BUSINESS	101	ARONSON, S.I.L.	84
			CARTER, R.A.S.	75
			MCKILLOP, L.G.	70
			MULDOON, B.L.	100
			ZINSZER, J.E.	55
	MATHEMATICS	102	BROWN, J.A.	77
			CARTER, R.A.S.	100
			KRIST, T.S.E.	70
			MCKILLOP, L.G.	26
			MULDOON, B.L.	79
			STROBEL, D.L.	60
			ZINSZER, J.E.	83

Figure 109: LIST BY SEMESTERNUM BY COURSENAME - part 2

PAGE 1

SEM	COURSENAME	NUM	INSTNAME	AVG GRADE
---	------------	---	--------------------	-----
F87	BUSINESS	101	HUTT, S.E.	81
		102	TROUSSLER, M.B.	77
	COMPUTER SCI	101	HILL, V.J.	70
	ENGLISH	101	PRIDE, J.G.	72
	MATHEMATICS	101	MORANSKI, J.W.	61
F88	BUSINESS	102	HUTT, S.E.	68
	COMPUTER SCI	102	ROBINSON, N.	89
	MATHEMATICS	101	JOHNSTON, J.G.	70
S88	BUSINESS	102	TROUSSLER, M.B.	76
	COMPUTER SCI	101	ROBINSON, N.	70
		102	ROBINSON, N.	75
	ENGLISH	101	PRIDE, J.G.	72
W88	BUSINESS	101	HUTT, S.E.	77
	MATHEMATICS	102	JOHNSTON, J.G.	71

Figure 110: NOMAD2 report showing average grades

PAGE 1

PROFDEPT	ASSOC AVG SALARY	ASST AVG SALARY	FULL AVG SALARY
------------	-------	-------	-------
BUSINESS	$58,000	N/A	$61,000
COMPUTER SCI	$55,000	N/A	$60,000
ELEC ENGRING	$54,100	$53,400	N/A
ENGLISH	N/A	$50,600	$59,000
FRENCH	$51,833	N/A	N/A
MATHEMATICS	$53,800	N/A	$58,000

Figure 111: NOMAD2 report with vertical and horizontal sequencing

```
                              PAGE    1

GRADE REPORT FOR STUDENT STROBEL, D.L.        ID:  1007

SEM  COURSENAME    NUM  GRADE
---  ------------  ---  -----
F87  BUSINESS      101     62
     COMPUTER SCI  101     88
     ENGLISH       101     68
F88  BUSINESS      102     60
     MATHEMATICS   101     80
S88  COMPUTER SCI  102     80
W88  MATHEMATICS   102     60
```

Figure 112: Grade report for a student

```
                                   PAGE    1

PROFNAME               PROFDEPT
--------------------   ------------
HILL, V.J.             COMPUTER SCI
MORANSKI, J.W.         MATHEMATICS
SAMPSON, R.B.          MATHEMATICS
PRIDE, J.G.            ENGLISH
BROWN, J.F.            FRENCH
HUTT, S.E.             BUSINESS
ROBINSON, N.           COMPUTER SCI
TROUSSLER, M.B.        BUSINESS
CRAIG, H.G.            ELEC ENGRING
DREW, S.C.             ELEC ENGRING
GREER, G.E.            FRENCH
WEXLER, R.L.           ENGLISH
JOHNSTON, J.G.         MATHEMATICS
CASSON, L.C.           FRENCH
```

Figure 113: LIST PROFNAME PROFDEPT

- *join* - combines two relations using a common field. In NOMAD2, a variation of this is accomplished by using the EXTRACT option of the DEFINE statement described earlier; it can also be done by using the MATCHING option of the LIST command.

  For example, Profid appears in the master PROFESSORS and within the master COURSES. The two entities may be merged in a report:

  LIST BY PROFID PROFNAME PROFRANK MERGE MATCHING -
  INSTNAME SEMESTERNUM COURSENAME COURSENO

  The *merge matching* operator is used to combine all data items contained in either or both masters (Figure 114). Notice that some professors have no associated courses and some courses are not associated with professors. To select data that appear in both masters only,

  LIST BY PROFNAME PROFRANK SUBSET MATCHING
  INSTNAME -
  SEMESTERNUM COURSENAME COURSENO (Figure 115)

- *difference* - given two relations A and B, the difference operator extracts instances appearing in A and not in B. A variation of this operator is available in NOMAD2: *reject matching* results in the selection of data that appear in the first master and not in the second. For example, to find all professors who were not instructing during the period fall 1987 through fall 1988,

  LIST BY PROFNAME REJECT MATCHING INSTNAME -
  TITLE 'PROFESSORS NOT TEACHING FROM F87 THROUGH
  F88' (Figure 116)

11.4.7 Other NOMAD2 Features

In this section, some additional features of NOMAD2 are briefly highlighted.

PAGE 1

PROFNAME	PROF RANK	SEM	COURSENAME	NUM
--------------------	-----	---	------------	---
BROWN, J.F.	ASSOC			
CASSON, L.C.	ASSOC			
CRAIG, H.G.	ASST			
DREW, S.C.	ASSOC			
GREER, G.E.	ASSOC			
HILL, V.J.	FULL	F87	COMPUTER SCI	101
HUTT, S.E.	FULL	F87	BUSINESS	101
		W88	BUSINESS	101
		F88	BUSINESS	102
JOHNSTON, J.G.	ASSOC	F88	MATHEMATICS	101
		W88	MATHEMATICS	102
MORANSKI, J.W.	FULL	F87	MATHEMATICS	101
PRIDE, J.G.	ASST	F87	ENGLISH	101
		S88	ENGLISH	101
ROBINSON, N.	ASSOC	S88	COMPUTER SCI	101
		F88	COMPUTER SCI	102
		S88	COMPUTER SCI	102
SAMPSON, R.B.	ASSOC			
TROUSSLER, M.B.	ASSOC	F87	BUSINESS	102
		S88	BUSINESS	102
WEXLER, R.L.	FULL			

Figure 114: Example of *merge matching*

11.4.7.1 System Variables, User Variables and Expressions

NOMAD2 supports the use of variables, as other programming languages do. Variables are prefixed by an ampersand (e.g., &DATE), and may be either system or user defined.

Some system ampersand variables (called system &variables) can be used to customize report formatting. For example, the variable &INDENT indicates the number of spaces to indent the left margin of a report. Its default value is zero, but it may be modified by a user for the duration of a terminal session. The following LIST command

```
                                                PAGE    1

                      PROF
PROFNAME              RANK   SEM  COURSENAME     NUM
--------------------  -----  ---  ------------   ---
HILL, V.J.            FULL   F87  COMPUTER SCI   101
HUTT, S.E.            FULL   F87  BUSINESS       101
JOHNSTON, J.G.        ASSOC  F88  MATHEMATICS    101
MORANSKI, J.W.        FULL   F87  MATHEMATICS    101
PRIDE, J.G.           ASST   F87  ENGLISH        101
ROBINSON, N.          ASSOC  S88  COMPUTER SCI   101
TROUSSLER, M.B.       ASSOC  F87  BUSINESS       102
```

Figure 115: Example of *subset matching*

```
                                              PAGE    1
                  BUDGET ANALYSIS

                                                % OF
FACNAME                  FACBUDGET              TOTAL
--------------------  ------------  -------------
ARTS                    $1,433,000          24.23
ENGINEERING             $2,580,000          43.63
SCIENCE                 $1,900,000          32.14
                      ============  =============
                        $5,913,000         100.00
```

Figure 116: Example of *reject matching*

```
&INDENT = 4 -
&TWIDTH = 80 -
LIST BY STUDMAJOR STUDNAME STUDID -
TITLE 'REPORT OF STUDENTS WITH THEIR MAJORS' &DATE
```

causes the report to be indented four spaces at the left-hand margin and sets the width of the terminal (&TWIDTH) to be eighty characters. Today's date will be printed as part of the title. There are many more system &variables, but they will not be discussed here.

Sometimes one needs to perform calculations that cannot be computed through the use of built-in functions. In such cases, one can define a variable (called a user &variable) and use the SET option of the LIST command.

Suppose a report showing the percentage of each faculty's budget compared to the total budget for all faculties is required. The command

```
LIST SUM(FACBUDGET) SET &TOTBUDGET NOPRINT -
BY FACNAME FACBUDGET (FACBUDGET/&TOTBUDGET*100) -
HEADING '% OF:TOTAL' -
TOTAL ALL -
TITLE 'BUDGET ANALYSIS'
```

will produce the report in Figure 117.

```
                                                    PAGE  1
                        BUDGET ANALYSIS

                                                      % OF
FACNAME                     FACBUDGET                TOTAL
--------------------    ------------   -------------
ARTS                      $1,433,000             24.23
ENGINEERING               $2,580,000             43.63
SCIENCE                   $1,900,000             32.14
                        ============   =============
                          $5,913,000            100.00
```

Figure 117: Calculations and reporting

The first line of the preceeding LIST command will cause NOMAD2 to sum the budget of each department and place the result into the variable &TOTBUDGET, without printing. The second line names the fields and expressions to be printed. The third line states that the expression (FACBUDGET/&TOTBUDGET*100) is to be called "% OF TOTAL." The fourth line indicates that all numeric fields are to be totalled, and the fifth line gives the title of the report.

As an example of a simple "what if" analysis, suppose a 5 percent salary increase for all professors is being considered. It is simple to produce a report showing current salaries, and what salaries would be if they were increased by 5 percent.

```
LIST BY PROFNAME PROFRANK PROFSALARY (PROFSALARY*1.05) -
AS $99,999  HEADING 'NEW:SALARY'
```

The resulting report is shown in Figure 118.

PAGE 1

PROFNAME	PROF RANK	SALARY	NEW SALARY
BROWN, J.F.	ASSOC	$52,400	$55,020
CASSON, L.C.	ASSOC	$54,300	$57,015
CRAIG, H.G.	ASST	$53,400	$56,070
DREW, S.C.	ASSOC	$54,100	$56,805
GREER, G.E.	ASSOC	$48,800	$51,240
HILL, V.J.	FULL	$60,000	$63,000
HUTT, S.E.	FULL	$61,000	$64,050
JOHNSTON, J.G.	ASSOC	$52,600	$55,230
MORANSKI, J.W.	FULL	$58,000	$60,900
PRIDE, J.G.	ASST	$50,600	$53,130
ROBINSON, N.	ASSOC	$55,000	$57,750
SAMPSON, R.B.	ASSOC	$55,000	$57,750
TROUSSLER, M.B.	ASSOC	$58,000	$60,900
WEXLER, R.L.	FULL	$59,000	$61,950

Figure 118: Report showing all salaries increased by five percent

The "AS $99,999" option causes the calculated new salary field to be printed as currency in whole dollars. Calculated expressions can be more complex than the above example. Suppose that full professors will receive 5 percent raises, associate professors will receive 4 percent raises, and all others will receive 6 percent raises. This can be accommodated by using an expression within the LIST command.

```
LIST BY PROFNAME PROFRANK PROFSALARY -
(IF PROFRANK='FULL' THEN PROFSALARY*1.05 ELSE -
IF PROFRANK='ASSOC' THEN PROFSALARY*1.04 ELSE -
PROFSALARY*1.06) -
AS $99,999 HEADING 'NEW:SALARY'
```

The report for such a command is shown in Figure 119.

PAGE 1

PROFNAME	PROF RANK	SALARY	NEW SALARY
BROWN, J.F.	ASSOC	$52,400	$54,496
CASSON, L.C.	ASSOC	$54,300	$56,472
CRAIG, H.G.	ASST	$53,400	$56,604
DREW, S.C.	ASSOC	$54,100	$56,264
GREER, G.E.	ASSOC	$48,800	$50,752
HILL, V.J.	FULL	$60,000	$63,000
HUTT, S.E.	FULL	$61,000	$64,050
JOHNSTON, J.G.	ASSOC	$52,600	$54,704
MORANSKI, J.W.	FULL	$58,000	$60,900
TROUSSLER, M.B.	ASSOC	$58,000	$60,320
PRIDE, J.G.	ASST	$50,600	$53,636
ROBINSON, N.	ASSOC	$55,000	$57,200
SAMPSON, R.B.	ASSOC	$55,000	$57,200
WEXLER, R.L.	FULL	$59,000	$61,950

Figure 119: Report showing differential salary increases

11.4.7.2 Procedures and the CREATE Command

Much of commercial data processing is concerned with the production of periodic reports, data entry and data updating. In NOMAD2, as with other programming languages, it is possible to store the necessary instructions so that such functions are executed automatically. Sets of NOMAD2 commands are called *procedures*, and can be created in a manner similar to the creation of a database schema. The desired commands are entered into a file using a text editor and saved. Then, to execute the commands, the name of the file is given, and the commands are automatically executed.

Another useful feature of the language is its ability to dynamically modify the database structure. This is done using the CREATE command. The LIST command is used for creating reports that can be displayed at the terminal or printed, but not stored. By using the CREATE command instead of LIST, a file and database are produced, which can be referred to by subsequent commands. A complete treatment of this topic can be found in McCracken, [4], and is beyond the scope of this text; nonetheless, it is a feature which the authors find noteworthy.

11.4.7.3 Fourth-Generation Languages: Summary

NOMAD2 is a relational database management system and application generator. It provides such functions as report and application generation, database creation and maintenance, graphics, statistics, and decision support. The development of reports and applications is provided by LIST and CREATE commands which extract data from existing databases, and compute, sort, format, print and create new databases where required. Within these commands normal computational functions, relational operations and report layout procedures are available. In addition there are statistics and graphics utilities plus what-if and goal-seeking functions.

The database management facilities include maintenance procedures for updating (insertion, changing and deletion), describing (schema), and protecting files. The most profound aspect of NOMAD2 and other fourth-generation languages is the fact that as DBMS's they are *self-describing*. This means that once a schema is created it provides the means for navigating through the database files. Thus, such systems are more than just organized sets of files; they provide powerful features for extracting and updating, all through their self-describing schema. Of course the query and application development languages are also significant, but their use would be very limited without the ease of navigation offered by the DBMS.

11.5 Data for the UNIVERSITY database

FACNAME	FACBUDGET
ARTS	$1,433,000
ENGINEERING	$2,580,000
SCIENCE	$1,900,000

Figure 120: FACULTIES data

DEPTNAME	DEPTBUDGET	DEPTFAC
BUSINESS	$4,200,000	ARTS
COMPUTER SCI	$3,210,000	SCIENCE
ELEC ENGRING	$3,005,000	ENGINEERING
ENGLISH	$2,690,000	ARTS
FRENCH	$1,450,000	ARTS
MATHEMATICS	$4,400,000	SCIENCE
MECH ENGRING	$2,742,000	ENGINEERING

Figure 121: DEPARTMENTS data

SEM
F87
F88
F89
S88
S98
W88
W89

Figure 122: SEMESTERS data

STUDID	STUDNAME	STUDMAJOR
1000	BROWN, J.A.	ENGLISH
1001	MULDOON, B.L.	MATHEMATICS
1002	ZINSZER, J.E.	COMPUTER SCI
1003	WELLER, M.E.	MATHEMATICS
1004	KRIST, T.S.E.	ENGLISH
1005	MCKILLOP, L.G.	BUSINESS
1006	CARTER, R.A.S.	COMPUTER SCI
1007	STROBEL, D.L.	BUSINESS
1008	BROOKS, F.	BUSINESS
1009	MORANSKI, R.	FRENCH
1010	DENT, C.L.	ENGLISH
1011	RAUCHWAY, A.E.	FRENCH
1012	ARONSON, S.I.L.	FRENCH
1013	SHORT, L.F.	COMPUTER SCI

Figure 123: STUDENTS data

EMPLOYEE NUMBER	PROFNAME	PROF RANK	SALARY	PROFDEPT
112-334	HILL, V.J.	FULL	$60,000	COMPUTER SCI
123-456	MORANSKI, J.W.	FULL	$58,000	MATHEMATICS
146-892	SAMPSON, R.B.	ASSOC	$55,000	MATHEMATICS
238-945	PRIDE, J.G.	ASST	$50,600	ENGLISH
390-800	BROWN, J.F.	ASSOC	$52,400	FRENCH
405-192	HUTT, S.E.	FULL	$61,000	BUSINESS
654-448	ROBINSON, N.	ASSOC	$55,000	COMPUTER SCI
667-890	TROUSSLER, M.B.	ASSOC	$58,000	BUSINESS
678-631	CRAIG, H.G.	ASST	$53,400	ELEC ENGRING
753-994	DREW, S.C.	ASSOC	$54,100	ELEC ENGRING
841-325	GREER, G.E.	ASSOC	$48,800	FRENCH
879-954	WEXLER, R.L.	FULL	$59,000	ENGLISH
977-320	JOHNSTON, J.G.	ASSOC	$52,600	MATHEMATICS
990-765	CASSON, L.C.	ASSOC	$54,300	FRENCH

Figure 124: PROFESSORS data

COURSENAME	NUM	COURSEDESC	SEM	INSTRUCTOR	STUD	GRADE
BUSINESS	101	BUSINESS ORGANIZATION	F87	405-192	1000	90
					1003	85
					1004	76
					1007	62
					1010	80
					1013	92
			W88	405-192	1001	100
					1002	55
					1005	70
					1006	75
					1012	84
BUSINESS	102	FUNCTIONAL AREAS OF BUSINESSES	F87	667-890	1000	88
					1003	75
					1002	73
					1006	69
					1010	81
			F88	405-192	1001	56
					1007	60
					1009	77
					1011	80
			S88	667-890	1004	79
					1012	84
					1013	65
COMPUTER SCI	101	INTRODUCTION TO COMPUTER USAGE	F87	112-334	1000	55
					1004	72
					1007	88
					1010	76
					1012	60
			S88	654-448	1004	68
					1012	72
					1013	70
COMPUTER SCI	102	COMPUTER PROGRAMMING	F88	654-448	1001	91
					1002	100
					1004	84
					1006	90
					1011	73
					1012	98
			S88	654-448	1000	69

					1003	77
					1007	80
ENGLISH	101	INTRODUCTION TO ESSAY WRITING	F87	238-945	1000	65
					1001	71
					1002	80
					1003	78
					1004	56
					1005	98
					1006	60
					1007	68
			S88	238-945	1008	86
					1009	70
					1010	89
					1011	91
					1012	63
					1013	35

COURSENAME	NUM	COURSEDESC	SEM	INSTRUCTOR	STUD	GRADE
MATHEMATICS	101	CALCULUS I	F87	123-456	1000	60
					1003	22
					1004	56
					1005	81
					1006	76
					1010	73
			F88	977-320	1001	45
					1002	87
					1007	80
					1009	66
MATHEMATICS	102	ALGEBRA I	W88	977-320	1000	77
					1001	79
					1002	83
					1004	70
					1005	26
					1006	100
					1007	60

Figure 125: COURSES, OFFERINGS, CLASSLIST data

11.6 Exercises

1. Are the differences between so-called third-generation programming languages and fourth-generation programming languages sufficient to warrant being called different generations, or are the fourth-generation technologies nothing more than a collection of buzzwords? Support your opinion with at least one reference from the library relevant to each generation.

2. Explain how second- and third-generation software have influenced the need for, or creation of, the traditional life cycle (e.g., correctness, quality, lines of code, etc.).

3. Draw an analogy between building a tall building using concrete poured into forms versus pre-cast concrete and third- versus fourth-generation languages.

4. The revised UNIVERSITY schema shown in Figure 106 contains two SEGMENT (internal groupings), one of which is keyed (OFFERINGS). How does the aspect of being keyed or not, relate to the policies of the university being modeled? (i.e., can a student take the same course twice)? What would the policy changes be if both segments (OFFERINGS and CLASSLIST) were keyed? What would the policy changes be if only CLASSLIST were keyed?

5. List some features of NOMAD2 which relieve a programmer of very tedious tasks.

6. The schema concept in NOMAD2 is quite important, going beyond seeing NOMAD2 as just a language. Elaborate on the significance of the *schema.*

7. What drawbacks are there to packaged software and special purpose languages?

11.7 Bibliography

1. Joyce, Edward J. *The Art of Space Software.* DATAMATION, Volume 31, No. 22., pp. 30-34. November 15, 1985.

2. Boar, Bernard H. *Application Prototyping: a Requirements Definition Strategy for the 80s.* New York, New York. John Wiley & Sons, 1984.

3. Martin, James. *Software Development without Programmers.* Englewood Cliffs, New Jersey. Prentice-Hall, 1984.

4. McCracken, Daniel D. *A Guide to NOMAD2 for Applications Development.* Reading, Massassachusetts. Addison-Wesley, 1980.

5. *NOMAD2 Reference Manual.* Wilton, Connecticut. D & B Computing Services, 1986.

6. Mosevich, J.W and McCandless, W. *Managing Fourth Generation Technologies: Guidelines for Action. CIPS/ACI Congress 86.* Vancouver, British Columbia. 1986.

Chapter 12

INFORMATION SYSTEMS PROTOTYPING

The previous two chapters have discussed some of the problems with traditional systems development and some of the tools developed to alleviate these problems. This chapter deals with the more popular alternatives to, and adaptations of, the traditional approach. These include prototyping, protocycling and end-user computing.

As the computing industry is continually evolving, one must keep in mind that fourth-generation languages and these new approaches are not the last word on systems development. However, they definitely have their places, and they have helped significantly in improving the productivity and quality of many software products. At the same time we must realize that their use does not preclude systems analysis or some equivalent, as presented in Parts 1 and 2. As we shall show, however, prototyping and protocycling can speed up the development process and handle poorly known or changing requirements much better than the traditional approach.

Our goal here is to present the commonly understood approaches to prototyping, protocycling and end-user computing. It is very important, and at times difficult, to be able to recognize where these approaches are applicable and where they are not. We will present some guidelines for applicability, and discuss the important skills and attitudes required by systems personnel for effective utilization of these approaches.

12.1 Prototyping

We are all familiar with prototypes of mechanical and other devices. These are normally the first versions of an object which are custom built, as opposed to mass produced, solely to test an idea or design. Most inventors make prototypes of their ideas and we often joke about these prototypes being "held together with bailing wire," as it would seem from photos of early aircraft, automobiles and computers.

In the context of software development we define a *prototype* to be a test version of a desired program or system, built quickly and not necessarily to any standards of good program design. The purpose of the prototype is to test an idea, or to verify and refine requirements, and to accomplish this very quickly. This is our own definition based upon our experiences. Other authors have their own versions of what is meant by a prototype, but they all mean virtually the same thing.

In practice there have emerged two basic implementations of the prototyping approach. One ends up by discarding the prototype and implementing in traditional languages, while in the other the prototype evolves into an end product. This latter approach, which we call *protocycling*, serves as an alternative to the traditional approach and is covered separately in the next section.

It should be obvious that prototyping is possible in COBOL, PL/1 or FORTRAN, but the tedium of setting up files and writing programs, which may be discarded later, makes these languages inappropriate. Clearly, fourth-generation languages are very well suited to prototyping and it is doubtful that prototyping would have developed as a separate method without them.

The simplest application of prototyping is one that aids in specifying requirements. Fourth-generation languages and screen-generating application development tools are very useful here. The typical situation is one in which the systems analysts and users are attempting to describe such things as complicated reports, screen output and input menus. Often users may know what they desire conceptually, but they cannot explain the details required for systems development. It is most effective for systems analysts to create their own interpretation, and show it to the users so that they have something concrete to work with. Now the users can either accept the proposed report, screen or menu, or specify how to change it. The process can be repeated until a final version is reached. An often-quoted saying attributed to users is "I don't know what I require. Show me something and I'll tell you if that is it."

Following is a simple example of a typical requirements specification problem solved by prototyping with NOMAD2.

A brokerage firm selling stocks and bonds to corporate clients with many accounts enters sales details into a batch COBOL system which, among many other things, produces reports showing monthly sales for each salesperson for each account. Management has requested a new report showing, for each salesperson, account number, the name of the corporate client associated with that account, profit, costs and other data. The purpose behind the requested report is to give management and each salesperson a picture of which corporate clients are good customers (generating lots of profit), which are poor, and which were once good but are now poor (therefore requiring attention).

Now if this request were implemented in COBOL as per the request "account number and associated corporate client," the manager would receive a report in a few weeks that would probably not be very useful. Why would it not be useful? Compare the reason behind the request to a sample output:

SALESPERSON	ACCOUNT	NAME	PROFIT
Jones	61124	Consol. Bank	$80,000
Smith	61772	Consol. Bank	$70,000
.	.	.	.
.	.	.	.
.	.	.	.

Assuming that each salesperson has around one or two hundred accounts and that corporate clients can have up to two hundred accounts, we see that while the report contains the information, it is of no use relative to its purpose. Quickly the systems analyst was aware of a possible ambiguity and decided to utilize NOMAD2 to prepare a prototype of the report as follows:

```
LIST BY SALESPER ACCOUNT -
CLNAME PROFIT
```

The analyst was able to deliver an *incorrect* report in only a few minutes. When shown the output, the manager pointed out that the information was wanted by client. The firm did not have to wait days or weeks to see the incorrect report, nor did they have to wait long for the required report shown below. (Note that each account is served by only one salesperson.)

LIST BY CLNAME BY ACCT PROFIT SALESPER

CLNAME	ACCT	PROFIT	SALESPER
Consolid. Bank	61124	$80,000	Jones
	61772	$70,000	Smith
	63744	$45,000	Craig
	64315	$90,000	Jones
.	.	.	.
.	.	.	.
.	.	.	.

When this *correct* report was shown to salespersonnel they pointed out its limitation – they would have to search the report for their clients. What they need is

LIST BY SALESPER BY CLNAME ACCT PROFIT

SALESPER	CLNAME	ACCT	PROFIT
Craig	Consolid. Bank	63744	$45,000
	.	.	.
	.	.	.
Jones	Consolid. Bank	61124	$80,000
		64315	$90,000
	.	.	.
	.	.	.
Smith	Consolid. Bank	61722	$70,000
	.	.	.
	.	.	.
.	.	.	.
.	.	.	.

And so we see that two reports were really required even though they were essentially the same. With the prototyping approach and the fourth-generation language, this problem was discovered and corrected very quickly and painlessly. It is not clear, one way or the other, if the need for two reports would have been identified using the traditional approach. The salespersonnel might not have seen the report until too late.

This is admittedly a very simple example, however, its lesson is that prototyping opens the way for experimentation and dialogue and, it is hoped, a better product.

The final implementation of those reports can now be made using the existing COBOL system where they are run regularly, or they can be run anytime with NOMAD2. Note that in the requirements specification type of prototyping, dummy data (or even no data) are often utilized. Thus no database is designed or files created.

Another type of prototyping occurs when the attributes of a proposed system are demonstrated, the system configuration experimented with and system feasibility verified. A typical case might be the following.

A market research analyst is asked by the president of the firm to access mainframe-based corporate historical data of sales, current and budgeted, and also the financial database of costs and revenues. The president requires a report and graphics of various related items and wants to be able to access the system through a menu-driven system on a personal computer. The theory here is that up-to-date financial data, together with historical data will allow senior managers to appraise the company "in real time," and to compare with past performances. The traditional approach would be to spend a few weeks analyzing and designing, followed by a few months programming and testing. There are, unfortunately, several unknowns here.

- How, precisely, does the president wish the reports and graphics to be presented?
- Once the president sees some output will it be the old story of "that's not exactly what I had in mind...?"
- Assuming it is known how to relate historical and current financial data, and that reports and graphs are well defined, what response time will be acceptable?
- Will there be technical problems with drawing graphs quickly on the personal computer, e.g., in data communication?

Clearly the systems division will be in trouble if it spends several months and lots of money to develop a system which is later not utilized. (This example is based on just such a case.) However, by prototyping quickly, and creating dummy data or realistic response times of dummy programs, with real screens and graphics, the president and/or systems people can appraise the expected realistic operation of the system as well as straighten out requirements. The technical problems of data communication and the use of a personal computer can also be evaluated. For example, will the president be satisfied with typing in commands or are function keys or a mouse best? Will the personal computer have a satisfactory response time and will data communication be a problem?

By prototyping in a simulation mode, many of these questions can be answered, and with little risk or investment. In the actual case upon which this example was based it was found that the best solution was not a personal computer, but rather a minicomputer linked to the corporate mainframe. The reasons were:

- Response time of the host-PC link was poor and inconsistent due to data communication problems, but more significantly due to the computational load on the host at peak hours.
- Other senior executives requested access to the system, putting additional pressure on the host-PC architecture.
- Much of the computation could be performed during the evening and loaded into the mini, giving consistent, fast responses.
- The PC graphics capacity was inferior to that of a mini.

In conclusion, prototyping for requirements specification and feasibility simulation is a very cost-effective method of systems analysis or design. It can give better-quality systems with better understanding between users and systems personnel. It can be effectively integrated into the traditional life cycle or it can replace traditional software development by applying it according to the technique outlined in the next section. To successfully apply prototyping, powerful fourth-generation tools are necessary (fourth-generation languages, with DBMS, spreadsheet, etc.). Also, the objectives and meaning of this approach must be clearly explained to users and understood by the systems staff.

12.2 Protocycling

Protocycling is a natural consequence of the prototyping philosophy and it is the power of the fourth-generation tools and modern hardware. Simply stated, protocycling iterates the prototyping method on the whole system being developed until a final acceptable version is obtained and the final prototype becomes the final system. That is, the prototype is not discarded, and no further third-generation or traditional programming is required. (Sometimes certain modules are programmed in assembler, COBOL or FORTRAN, but they represent only a small part of the final product).

There are no clear-cut rules for judging when protocycling is applicable but experience indicates some typical cases:

- Requirements are not well understood.
- Requirements are expected to change fairly often due to external factors (e.g., tax law changes).
- The proposed system must be developed very quickly.
- The proposed system is an implementation of an untried or theoretical business process.

Notice how these points closely match those where fourth-generation tools are recommended.

Areas where protocycling (but not necessarily prototyping) and fourth-generation languages are *not* appropriate include:

- systems requiring extremely fast response times
- systems with static, well-defined requirements
- implementations on hardware with non-standard operating systems or where no fourth-generation tools are available.

Thus, we do not propose that all systems be developed by the methods and with the tools described in Part 3.

In the case of a protocycling implementation we can reconfigure the life cycle as shown in Figure 126. Notice the correction of the database structure. This is of great importance because the database structure will influence ease of change to code, understandability and enhancibility of the system by analysts and users, and efficiency of the system as a whole.

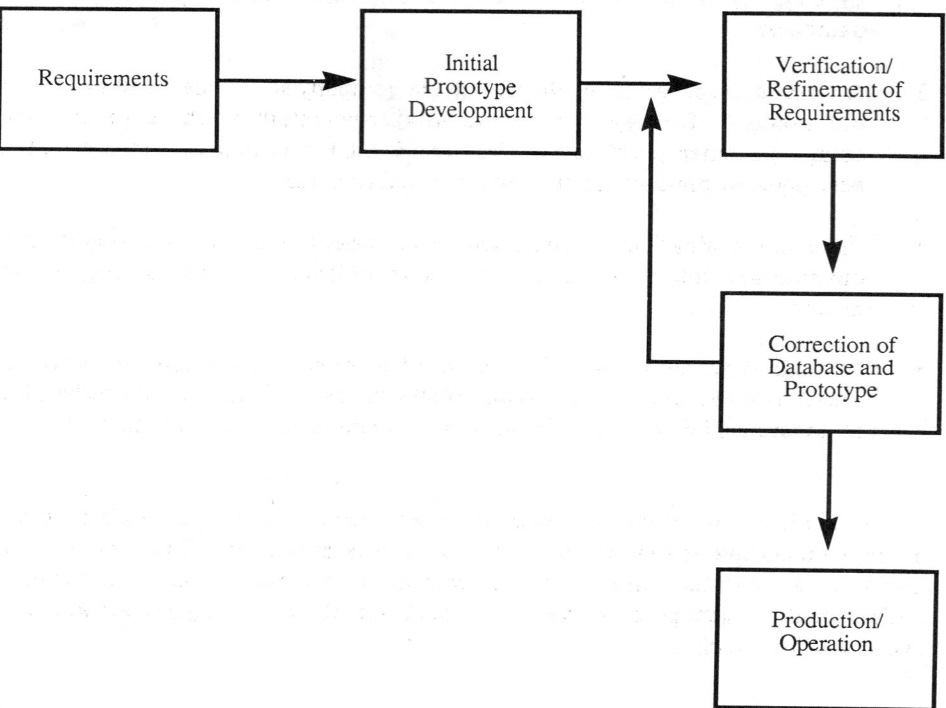

Figure 126: Protocycling

A very successful protocycling project for a major corporation is now discussed.

The corporation in question perceived a need to totally change its marketing strategies. A radically new approach was conceived whereby:

1. Historical data on customers was to be made available in a database. This would indicate what products were purchased, when, by whom, and their

reasons (for private or business purposes, etc.). This data would include the company's own products and those of its competitors.

2. Economic data and demographic data and market research, together with the history database, would be utilized to forecast demand for products, by city, region and even country. This would create a potential-market database.

3. A simulation program would match the company's and competitors' potential product offerings against the market-potential database to evaluate actual customer purchases. This would allow the company to create the best possible product (including price, quality, etc.).

4. Financial evaluation of company costs, prices and volumes would give quantitative dollar evaluation of product offerings, by city or region and in total.

5. Marketing personnel would be required to work in an automated environment, utilizing personal computer workstations, and would need to be able to generate their own applications in a fourth-generation environment.

A detailed requirements specification was developed and a rough estimate of implementation resources and time frame was produced. The total cost in human and computer resources was estimated in the low millions, and the time to first module turnup at two years. In addition, there were many serious risks and unknowns, such as

- Is it really possible to forecast demand?
- Is simulation of customer behavior possible?
- Can the historical, economic and demographic data be obtained in a timely fashion?
- Individually the separate ideas (1 to 5) are possible, subject to the above questions, but will there be the required synergy when the total system is integrated? Will one weak link jeopardize the whole idea?

- Will the marketing people desire, or be able, to work in an automated environment?

It would have been very risky indeed to assume that this approach would work well enough for the corporation to spend two years and rely on an untried concept. During those two years the competition might try something similar and certainly the whole environment would change. Therefore it was decided to prototype the concept, utilizing a fourth-generation language together with commercially available forecasting and simulation packages.

The initial trial was configured as follows:

Several personal computer workstations were purchased and linked to the corporate mainframe. A fourth-generation language was installed on the mainframe as was its equivalent on the personal computers. Also, the workstations were equipped with electronic spreadsheet, graphics and word processing software. While the prototype was being set up, the project team made presentations to the entire marketing division about the project and its goals and about prototyping and fourth-generation languages.

The goals were

- to assess the feasibility of potential market projection, by cities, regions and various geographic areas
- to use simulation
- to automate the environment.

The first step was not too difficult: government data were available about the historical purchases of products, although they were six months old. More difficult was how to handle the forecasting and economic indicator data. Figures such as projected GNP and the consumer price index are uncertain with regard to their performance. It was decided that for the prototype, with goals of testing feasibility, *subjective* forecasting would be adequate. This meant that the marketing person would only have to adjust historical figures (upward or downward) to obtain a projection.

The fourth-generation language was used in the following ways:

- A database of historical data was created and the fourth-generation language was used to create an information system. Marketing analysts could query the database themselves and a selection of queries was also made available.
- It was possible to link the simulator to the historical and forecast databases as well as to produce regular and customized reports.
- Experimentation with financial evaluation reports and graphs of many variables was performed.
- A simple-to-use interface made possible the use of spreadsheet software for further analysis of output of both the information system and simulator.

During the first year of operation the prototype was able to function as a small-scale production system. Its value was proven when several urgent questions posed by senior managers were answered in record time, for example, "Why aren't we selling product X in region Y?" and the answer: "The competition are killing us there with their product Z." The historical database was redesigned about five times to better accommodate the types of queries and programs requested. It was then considered stable. The personal computer population grew very quickly due to demand by other personnel.

The lessons learned from prototyping were as follows:

1. Marketing personnel desired and even needed to work in an automated environment.
2. The availability of a market history information system was very valuable.
3. The simulator was quite useful, not just because it predicted precise purchasing figures. It was also useful because when it predicted surprising results, it forced its users to ask "now why does it say this when we have been thinking that?" In other words it forced and helped the understanding of specific markets.

4. Marketing personnel were quite happy to create different scenarios and to forecast by themselves. They decided to abandon, for the time being, any use of sophisticated forecasting systems.

5. The financial evaluator ultimately developed was very useful, especially when presenting new products and strategies to senior management.

6. Because marketing personnel worked on a daily basis with systems people, a better product was developed, and more importantly, the marketing people and systems people developed a better understanding of each other's jobs and environments.

The prototyped system evolved and grew over an eighteen-month period to a point where it was considered a full production system. Thus it was really a protocycling approach in the sense defined above. Because CPU resources and response times were within reason, no one ever suggested redesigning and reprogramming in a third-generation language. Both the code and design were thus created in an iterative fashion, not sequentially as in the traditional approach. Also, requirements were created, changed and implemented for a few iterations and they soon became stable.

Some systems management problems did surface and required their own experimentation. The main one concerned maintenance. Many companies separate development and maintenance. A protocycled system is more like a dynamic or never-finished product, there being no distinct phases. The agreed-upon approach was to assign a systems team to a department (in this case marketing) and they were responsible for all aspects of their protocycled products. No maintenance phase was established.

Another systems management problem concerned the need to deal continually with users. Programmers are used to working alone, with terminals, and not sitting next to their customers. This problem was initially solved by choosing those systems people known to have good interpersonal skills. Fortunately many such people exist, some with latent skill that only needed the exposure.

Protocycling, then, is in some ways a more ideal mode of systems development than the usual specify-everything sequential mode, for it better reflects the evolutionary nature of software (i.e., it views software more as a living entity than as a dead entity).

12.3 End-User Computing and the Information Center

A parallel development to prototyping is end-user computing. Both require the power and relative ease of use of fourth-generation tools. As the name implies, *end-user computing* is the provision of computing facilities with which non-systems people can generate their own applications. The most common such facility is provided by personal computers with software such as electronic spreadsheets, business graphics and database management systems (with query languages). Fourth-generation languages, such as FOCUS, NOMAD2 and RAMIS, greatly expand the capabilities, on both mainframes and personal computers, far beyond that possible with COBOL or FORTRAN.

The provision of end-user computing in most companies is through the concept of an *information center*. This concept was developed in 1979 at IBM Canada, and it provides for corporate and some public information being made available to all IBM personnel. The information center concept has come to be interpreted as "any corporate facility for access to data and application generation capabilities." Thus an information center is a concept and not necessarily a physical place where users go to obtain information (although this is the case in some companies).

Currently, many companies have information centers, usually consisting of one or more departments in which staff provide training in fourth-generation products, and personal and mainframe computers, and offer advice on hardware and software acquisition. They do not themselves program for users, but rather they help users help themselves and in doing so they support end-user computing.

Successful information centers, that is, ones that persist and are of real value, seem to have one thing in common: staff who know the business and are able to listen and explain things well. To be effective the staff must be able to understand the corporate data and the meaning of that data to the company.

Returning to the subject of end-user computing, the most common form is obviously the one made possible by personal computers and associated packaged software (e.g., spreadsheets), and fourth-generation tools. Many users purchase their computers with no advice or involvement from systems, and they normally receive no support from the systems area, although they do from the information center. The types of applications created on personal computers are usually not complicated enough to require a systems analysis, although they may need a fair amount of preliminary thinking. Therefore, systems analysts are seldom directly involved with this type of application.

The next most common type of end-user computing does frequently require the help of the systems division, often of a systems analyst. This is the case where fourth-generation or special purpose languages are utilized by end-users themselves to create unique applications. The most critical problem to be faced is often understanding the corporate database. Where can one find the data needed for this application? If the data are not easily understood, or not readily accessible, then an analyst, and perhaps a database designer, will need to understand the requirement and create, or help create, a program or programs to extract the required data.

End-user computing, in its many forms, is on the increase. On the one hand, it does relieve the systems group of some development work. It also can increase resource consumption of mainframes and certainly has accounted for a large increase in the number of personal computers in major organizations.

12.4 Changing Skill Emphasis

The increased use of prototyping, protocycling and end-user computing has changed the relationship between systems people, especially analysts, and users. It is now common for the analyst to sit at a terminal with the user and together create a screen or write a program in a fourth-generation language. More than ever before it is necessary to develop

- sound knowledge of the business
- listening skills, diplomacy and patience (i.e., interpersonal skills).

The analyst will now need to work more with the user as a partner, to mutually solve business problems. This means that each must strive to better understand the other's environment and job, and in so doing, the gap between user and systems people will shrink [7].

Working more closely together will also require a kind of teaching on the part of the analyst, and the development of better interpersonal skills. For example, rather than tell a user that an idea will not work, the analyst must, through example, diplomatically show the user that the idea is impossible.

Analysts, systems people and users are becoming less separate, by business necessity and by the nature of the technology.

12.5 Conclusions

The tools and approaches covered in Chapters 11 and 12 are just some of the current trends in computing. Those presented here have been chosen because they are practised, not because they are elegant or theoretical. The automation of computing is probably going to lead to more progress in the automation of other areas, areas which we cannot even guess. For now, we see great use of fourth-generation tools and appropriate methods.

12.6 Exercises

1. Ad hoc, or one-time programs are often requested by management. List some possible cases where these might be required. Could these ad hoc management needs be met with traditional approaches? How does the information center help in a situation such as this?

2. Why are prototyping and protocycling fairly recent methods?

3. Is there any tangible value to prototyping? (Hint: cost of change in user requirements.)

4. What problems can you envision with protocycling?

5. What danger is there in staffing an information center with ″computer jockeys″?

12.7 Bibliography

1. Boar, Bernard H. *Application Prototyping: a Requirements Definition Strategy for the 80s*. New York, New York. John Wiley & Sons, 1984.

2. Budde, R., Kuhlenkamp, K., Mathiassen, L. and Zullighoven, H. (editors). *Approaches to Prototyping*. Berlin, West Germany. Springer-Verlag, 1984.

3. Carey, T.T. and Mason, R.E.A. *Information System Prototyping: Techniques, Tools, and Methodologies*. INFOR, Volume 21 Number 3. August, 1983.

4. Cash, James I., Jr., McFarlan, F. Warren and McKenney, James L. *Corporate Information Systems Management: Text and Cases*. Homewood, Illinois. Richard D. Irwin, 1983.

5. Martin, James. *An Information Systems Manifesto*. Englewood Cliffs, New Jersey. Prentice-Hall, 1985.

6. Martin, James. *Managing the Data Base Environment*. Englewood Cliffs, New Jersey. Prentice-Hall, 1983.

7. Martin, James. *Software Development without Programmers*. Englewood Cliffs, New Jersey. Prentice-Hall, 1984.

8. McFarlan, F. Warren and McKenny, James L. *Corporate Information Systems Management: The Issues Facing Senior Executives*. Homewood, Illinois. Richard D. Irwin, 1983.

9. McLeod, Raymond, Jr. *Decision Support Software for the IBM Personal Computer*. Toronto, Ontario. Science Research Associates, Inc., 1985.

10. Mosevich, J.W and McCandless, W. *Managing Fourth Generation Technologies: Guidelines for Action. CIPS/ACI Congress 86*. Vancouver, British Columbia. 1986.

11. Myers, Glenford. *The Art of Software Testing*. New York, New York. John Wiley and Sons, 1979.

12. Rockart, John F. and Bullen, Christine V. (editors). *The Rise of Managerial Computing: the Best of the Center for Information Systems Research*. Homewood, Illinois. Dow Jones-Irwin, 1986.

13. Sprague, R.H. and Carlson, E.D. *Building Effective Decision Support Systems*. Englewood Cliffs, New Jersey. Prentice-Hall, 1982.

14. Sprague, Ralph and McNurlin, Barbara. *Information Systems Management in Practice*. Englewood Cliffs, New Jersey. Prentice-Hall, 1986.

Appendix A

GUIDELINES FOR A STUDENT PROJECT

One of the most effective means by which systems analysis can be learned is through practice. In a course, this means a student project with teams of from two to four students. For the course taught by the authors to students in their final year of study, such student projects comprise roughly forty percent of the final grade. Over the past few years, the reputation of the student projects has spread throughout the business community, and we now have a waiting list of companies who wish to have a business analysis performed. Several student project groups have formed consulting firms as a result of this course.

The ideal maximum class size for such an undertaking is around eighty, although projects are feasible for larger class sizes. In the past, the authors have relied on a teaching assistant, with previous experience in the course, to administer and be responsible for the projects.

Each team is required to contact a local company of any size or a university or college department. In larger companies, the contact can be made through the systems department. For smaller companies and university and college departments a manager/owner or department head, respectively, is the best choice. Such a "cold call" is difficult, but it constitutes a good test of interpersonal skills.

Once it is established who to contact in the company or department an appointment is made over the phone. Further contact will be in person. During the initial face-to-face interview, contacts should be given a very brief verbal overview of system analysis if they do not already know what it is, and a description of the purpose of the student project. That purpose is to analyze (do not say study - this sounds too passive) and understand a limited business or business area. The view is to appraise potential computer support. Because the team is proposing this analysis out of the blue, the exercise is really a combination of requirements determination and systems analysis.

It must be made clear to the contact person that no programming of any type will occur: it is an *analysis* and conceptualization exercise. However, many projects undertaken by the authors' students for smaller companies have been implemented.

The end result will be a recommended systems approach if one is warranted. One student project undertaken for a roofing contractor recommended that the owner take a course in time management and purchase an electric typewriter. The students who performed this analysis exercised great diplomacy and tact when giving their results to the owner. They suggested a revision of manual procedures as well. Ultimately, the contractor was pleased and the team received very high marks.

The scope of the student project, then, is a *modest* business area. It is better to be too small than too large. The objective is a real systems analysis, complete with a conceptual data model, logical database schema, data flow diagrams or prototype, and data dictionary. As well, there must be a cost/benefit analysis and recommended automation strategy.

Within one week of having visited the company and obtaining company agreement to do the project, an initial report should be submitted to the course instructor, with a copy to the contact person within the targeted organization. This report, of one or two pages, should contain the following:

- members of the group
- company or organization for whom the analysis is being performed
- one- or two-paragraph description of the organization
- contact in the organization: name, address, telephone number.

The purpose of the initial report is to provide an indication of the scope of a potential project. The teaching assistant should provide relevant comments, and indicate suitability regarding the organization and scope of the potential project.

During the next three to four weeks, a progress report should be submitted giving the objective of the project, and the results of data analysis. Subsequent to the evaluation and revision of this report, a report containing the results of functional analysis (set of DFD's) or, as an alternative, evidence of a prototype built in a 4GL or DBMS, is required.

At the end of term, a final report is submitted. It is evaluated on the basis of the following criteria:

- format and clarity of presentation
- completeness, and consistency of data and function analysis
- understanding of the proposed systems and the current situation
- detection of problems within the current system
- cost and other justification for solutions proposed to solve problems.

Material that may be included in the final report follows.

1. *ABSTRACT*

   Executive summary (one page) indicating business requirements, systems requirements and recommended solution.

2. *INTRODUCTION*

   - brief description of the organization, its major activities and objectives
   - brief overview of the system being investigated
   - justification of analysis

3. *ANALYSIS*

- data analysis, function analysis or prototype
- cost/benefit analysis, budget and schedule
- data dictionary (optional)
- other documents generated during the project, such as physical DFD's and automation strategies

For the last assignment of the term, students may be required to present their projects to the class. Evaluation of presentations includes not only the projects themselves, but the quality of the presentation. For the course taught by the authors, students are required to evaluate each others' presentations as well as to present their own project as the last assignment of the term. Upon completion of the term, the authors have found it beneficial to thank the companies involved. As well, a list of "analyzed" organizations is maintained, and distributed to students at the beginning of each term.

Obviously, the benefits of a student project go far beyond the development of technical skills. Participants learn about group dynamics and company politics, and test their interpersonal abilities.

Following are a sample course outline and sample presentation evaluation forms.

A.1 Course Outline

A.1.1 Course Topics

1. The Business Context for Information Systems
2. Data Analysis
3. Data Modeling
   - conceptual modeling
   - logical modeling
4. Data Management
5. Function Analysis
   - dataflow diagrams
   - decision tables
   - decision trees
6. The Process of Analysis
7. Cost/Benefit Analysis
8. Presentation of Student Projects

A.1.2 Course Evaluation

1. Assignments - 30%
   - essay (typically based on a Harvard Business School case)
   - data analysis
   - function analysis

- presentation of student projects

2. Group Systems Analysis Project - 30%

3. Final Examination - 40%

A.2 Group Presentation Evaluation Form

Evaluator: Project Number:

Note: for rankings, 1 = very poor, 5 = outstanding. Justify each ranking in the space allotted.

1. Rank overall organization (includes flow between speakers, division of subject matter, effectiveness of wrap-up)

   1 2 3 4 5

   major strengths:

   aspect(s) needing improvement:

2. Rank use of audio-visual aids and handouts

   1 2 3 4 5

   major strengths:

   aspect(s) needing improvement:

3. Rank credibility (includes accuracy of assertions and usage of data)

   1 2 3 4 5

   justification:

4. Rank the group's overall delivery

1 2 3 4 5

major strengths:

aspect(s) needing improvement:

5. How clearly/effectively was the overview of the business, and the area under investigation described?

1 2 3 4 5

major strengths:

aspect(s) needing improvement:

A.3 Individual Speaker Evaluation

Evaluator: Speaker: Project Number:

1. Rank organization and flow

1 2 3 4 5

major strengths:

aspect(s) needing improvement:

2. Rank use of body language (eye contact, gestures, posture) and projection to audience (tone of voice, choice of words, confidence, enthusiasm)

1 2 3 4 5

major strengths:

aspect(s) needing improvement:

3. Rank use of audio-visual aids or handouts

1 2 3 4 5

major strengths:

aspect(s) needing improvement:

4. Rank the credibility

1 2 3 4 5

major strengths:

aspect(s) needing improvement:

5. Additional comments

Appendix B

CASE STUDIES

B.1 Quality Textiles International

This case involves a wholesale textile company for which research was done by the authors. It provides a good example of current automation needs and trends. The case is especially suited for assignments regarding E-R modeling, relational database design, DFD's, and prototyping (which was done by one of the authors).

The manual system works well, however it is calculation intensive and does not provide higher-level management information. It is interesting to note that in general, the inventory control function of the textile industry has not been successfully automated. Not only may many currencies be involved in calculating actual landed costs of products, but there are also issues regarding quality: the number of flaws and where they occur within a roll of fabric, color matching, etc. These are problems that have not been addressed by automation.

B.1.1 Background

The company involved in this case is in the wholesale textile business, supplying fabrics to the furniture trade. Its customers include furniture manufacturers; architects; designers employed under contract to major corporations, hotel chains, government, etc.; upholstery shops; and retail furniture stores. Most of the fabrics sold by the company are imported from Europe. The company purchases from mills all over the world, and currently deals in twelve to twenty foreign currencies. The company has showrooms and agents across the country, but their only warehouse is located at the corporate head office.

The costing function of the inventory has two aspects: costing to arrive at an actual landed cost per yard for each roll in a shipment of fabrics kept in inventory, and a costing procedure to produce a price list for resale of these fabrics. Fabric is sold by the yard, and in the inventory records each roll is referred to as a *piece*. The company purchases not by the yard or meter, but by the piece. There is no standard length for a piece, but usually they are about 40 to 55 yards in length.

The costing to arrive at actual landed cost is a complicated process for several reasons. First of all, the company purchases in the originating currency, whose exchange rate is subject to fluctuations. When a shipment arrives at the local customs office, if the fabric is dutiable, duty is payable at the particular exchange rate of the day it clears customs. (The conversion rate used by customs is always a bit lower than the quoted bank rate, because of the built-in profit of the bank's rate.) The company may pay the invoice 30, 60 or 90 days later, and in the interim, the conversion rate may have changed.

When calculating the landed cost, the company uses the conversion rate furnished by the company's bank for the date the goods arrive at the company's warehouse. This date could be one or two days after clearance by customs, however, it is used because this is the earliest date the company could consider paying for the goods just received. Management believe they cannot wait until the invoice is paid when calculating landed cost.

Several different duty rates apply to the company's products, and each relates to the content of the fabrics. Of course, one shipment can include several kinds of fabrics. (See B1.6 for a list of duties.) The duties are factors that are applied to the yardage indicated on the customs slip.

The company's suppliers fill out the required customs slips. Suppliers normally measure fabric while it is still on the loom, and customs slips are completed using this measurement. However, when the fabric is removed from the loom, there is usually some shrinkage, and the company may receive less yardage than the original mill measurement. The company measures each roll and allows up to 5 percent shrinkage allowance. Goods with more than 5 percent shrinkage are returned to the supplier. There are also times when the fabric has flaws in it. The mill normally gives an allowance for this – extra yardage for which there is no charge or duty. The company examines each roll for flaws, and if the usable yardage and the mill's usable measurement differ by 5 percent or less, the goods are accepted.

B.1.2 Costing Shipments of Fabrics

Exhibits A1 through A4 are three invoices from a supplier plus a page of customs documentation. Notice on exhibit A1 that "quantity shipped" is 53 yards but that 52.5 yards were actually received. The difference is the shrinkage. Occasionally the company receives a bit more actual yardage than was ordered. Currently the company does not keep track of variance between the mill's measurement and usable yardage, but would like this information as part of a management information system. It would be useful to have information such as, what the difference was for total usable yardage versus suppliers' measurements for the past year or for a particular supplier for a specified time period.

Each piece is given a unique number by the supplying mill, and even though two pieces may be identical with respect to pattern and color, each can be identified by number. The company uses these same numbers to identify each piece in inventory. (This numbering system also allows the landed cost to be specific to each piece in stock.) Freight costs may vary with each shipment, and for small shipments received, the freight costs are proportionally higher than for larger shipments.

There are generally two categories of customers: the first is manufacturers and wholesalers, and the second, retailers (department and furniture stores, interior designers). The markups are different for each category.

Each pattern and color within a pattern is numbered by the mills and assigned a name by the company.

Exhibits B1, B2, C1 and C2 are two entries from the company's costing book, called "the black book," which is used for pricing the company's products. Information for each pattern is given on two sides of an 8 1/2 x 11 sheet of paper. Information relating to the fabric as a whole appears on the front (exhibits B1 and C1):

- the mill's pattern number - the company's pattern name
- the current mill price
- the mill's color numbers - the company's color names,

- the supplier's name and address
- terms
- possible shippers.

On the back (exhibits B2 and C2), there is other information regarding the fabric, such as

- type of fabric
- yarn content (which provides the basis for assessing duty)
- duty rate
- pattern repeat (length between pattern repetition).

The lower portion displays costing and pricing data for the fabric. Typically an entry is made in preparation for a new price list. The information includes

- DATE, date of the entry
- COST, cost (in originating currency) per yard
- RATE, current exchange rate
- OUR COST, the company's cost, which is COST x RATE
- DUTY, duty (per yard) based on the duty rate listed at the top of the page
- FRT, the freight (per yard), which is based on the average (or management's judgement) of the freight for shipments received during the period of the previous price list. The freight (factor) per yard is different for each supplier, but constant for all fabrics from the same supplier.
- LANDS, the landed cost.

Below the landed cost entry is a double entry, containing pricing information. One entry is for B-type customers (manufacturers) who have a lower markup; D-type customers (upholsters, designers, retailers) are charged a higher rate. At the right side of each entry is the new price and the markup used for getting from cost to price.

Exhibits B1 and B2 are the entries for the pattern, Allegro. The last entry for Allegro was done on August 22. The cost factors are as follows:

- DATE is Aug 22
- COST is 15.04 (UK) per yard
- RATE is 1.75
- OUR COST is 26.32 (15.04 x 1.75)
- DUTY is 6.58 (26.32 x .25)
- FRT is 1.60, and
- LANDS (landed cost) is 26.32 + 6.58 + 1.60 = 34.50.

The new B price is $69.00, a markup of 1.00 (100 percent). The multiplier to get from cost to price is 2.00, and the results are rounded. The new D price will be $78.00 (per yard), a markup of 1.4. So, the multiplier for the D entry is 2.4, and the results are again rounded. In general, each fabric is examined individually, and the markups are unique.

Exhibits D1 and D2 show two pages from the company's log book, which records shipments of goods upon arrival. For each shipment, the following information is given:

- CODE, the code number supplied by the company to each shipment
- DATE, date of arrival at the company's warehouse
- SUPPLIER, supplier from whom the goods have been received
- PATTERN, the company's pattern name (there may be several per shipment)
- COST PER Y, the cost per yard in the originating currency

- EXCH RATE, the current bank exchange rate
- OUR COST, the converted cost per yard
- DUTY RATE, the duty rate
- DUTY COST PER Y, the actual cost of duty per yard. This cost is calculated by dividing the duty paid to customs by the actual yardage received. (The actual yardage received shown in boldface type in exhibits A1, A2 and A3.) For example, Exhibit A2 shows 53.9 yards of Allegro, color Aurora received. The duty for this is $343.50, and it is the second entry on Exhibit A4, the customs importation form. Then the actual duty per yard for this shipment and fabric (part of shipment #314) is $343.50 / 53.9 = $6.37.
- TOTAL YD's RECD, total yards received this shipment (broken down per pattern)
- TOTAL FRT COST, the total freight cost for this shipment (this entry may be delayed by two weeks until the freight bill arrives)
- FRT COST PER Y, the freight cost per yard for this shipment (= TOTAL YD's RECD/TOTAL FRT COST)
- LANDED COST/Y NO FRT, the landed cost per yard without freight, (= OUR COST + DUTY COST PER Y)
- ACTUAL LANDED COST, the landed cost per yard including freight (= LANDED COST/Y NO FRT + FRT COST PER Y)
- BL BOOK LANDED, which shows the landed cost according to the black book
- BL BOOK FRT, the latest black book entry for the freight-per-yard cost.

Notice the "x" beside each entry in between the ACTUAL LANDED COST column and the BL BOOK LANDED COST column. An "x" is entered there if the actual landed cost is higher than the black book landed cost. When most entries contain an "x," it is time to prepare a new price list because the costing of current shipments is higher than the black book standard cost.

By re-examining exhibits B1 and B2, the black book page for Allegro, one sees that the latest FRT entry (freight cost per yard) is 1.60. By referring again to shipment code #314 in Exhibits D1 and D2, one can track actual shipment costs of the fabric Allegro, along with others in the shipment:

- COST PER Y (UK cost/yard), 15.04
- EXCH RATE, 1.80
- OUR COST, the converted cost per yard (15.04 x 1.80 = 27.07)
- DUTY RATE, 25%
- DUTY COST PER Y, derived from the customs invoice ($343.50/53.9 = 6.37)
- TOTAL YD's RECD, 53.9 yards for Allegro and 210.2 yards for the entire shipment
- TOTAL FRT COST, total freight cost of $437.20
- FRT COST PER Y, $437.20/210.2 = 2.07
- LANDED COST/Y NO FRT, 27.07 + 6.37 = $33.44 and
- ACTUAL LANDED COST, 33.44 + 2.07 = $35.51.

Notice the "x" beside the actual landed cost figure. This appears because the actual landed cost is higher than the black book standard of $34.50. In fact, all the entries have x's, an indication that a new price list is required.

The only remaining points we need to cover are the day-to-day aspect of inventory control and the physical taking of inventory.

B.1.3 Day-to-Day Inventory Control

After the pieces in a shipment have been examined and measured, they are ready to enter into inventory. The shipment has been assigned a code number and entered into the log, as discussed above. The warehouse foreperson fills out an inventory card in duplicate for each piece. Exhibit E shows an inventory card for the pattern Allegro in the color Aurora. The piece number is 739 and the shipment code is 314. One copy is kept in a bank of cards and is the responsibility of the foreperson. This card bank is ordered alphabetically by color within pattern. The other copy of the card is kept with the piece.

When a customer order is received, one of the staff fills out a cutting order (Exhibit F) which states the pattern, color and yardage ordered by the customer. This is given to the foreperson who examines the appropriate card or cards in the card bank and gives instructions to one of the cutters to cut a specific piece or pieces to fill the order. The piece is taken from inventory and cut. The cutter records the date, customer name, amount cut and flaw allowance (if any) on the card located with the piece and brings the card to the foreperson who updates the card bank. The total yardage, piece number(s) and allowance are entered on the cutting order. This system was set up by the company's auditors and works well.

After the order has been cut, the cutting order returns to the office and the office staff update their own set of records. These records keep track of incoming shipments and outgoing orders; their purpose is to give the office staff an idea of what is in stock, and enable the staff to be well informed when customers make telephone enquiries. (For accounting purposes, this set of records is completely redundant because it duplicates the inventory records in the warehouse.)

B.1.4 Quarterly and Year End Inventory Control

Inventory is physically counted quarterly and at fiscal year end. Exhibits G1 and G2 are sheets used for taking inventory. G1 is a detailed sheet for Allegro, and was used at fiscal year end. Several pieces for each color of Allegro were in stock. The yardage for each piece, unit (yard) cost and value are shown. The total value of Allegro on hand at year end was $4,740.14. Information for each entry is obtained from the card bank, and spot checks are performed at each inventory taking.

G2 is a summary sheet. It contains one entry per pattern, consisting of the following:

- PATTERN, the pattern name
- YARDS, yards in stock

- VALUE, value from the detailed inventory sheets
- MARKED DOWN VALUE, marked down value (if any) and
- AVG. VALUE PER YARD (= VALUE/YARDS)

This manual system works well, even though it is calculation intensive. One of the main benefits of automation, in addition to relieving the staff and management of the calculations, should be to provide information that until now has been unavailable.

B.1.5 Foreign Currencies

For the foreign currencies, use a listing that appears in the financial section of the newspaper.

B.1.6 Duties (may change annually)

FABRIC	*RATE*
cut pile, 100% cotton	15%
100% manmade fibers	25%
100% wool or hair	25%
woven, 100% cotton	17.9%
wool	25%
woven fabrics with cut pile,	
wholly or partially manmade fibers, no wool or hair	25%
knitted fabrics with cotton, synthetics, wool	25.4%
imitation suede	10%
silk	0%

B.1.7 Possible Assignments

1. Form teams of three, where two members portray management of the textile company described above and the third member portrays a systems analyst. Together develop a conceptual model of the company's data requirements, and document it using the E-R method.

2. Using the conceptual data model as the basis, design a relational database for Quality Textiles, and document it using data structure diagrams.

3. Draw a complete set of physical and logical data flow diagrams for the textile wholesaler described above. The physical diagrams should have at least three levels, and the logical diagrams at least five levels. Supply any missing elements, and note your assumptions. (Suitable for groups of two to three.)

4. Develop a prototype for the database, and for the functions of inventory control, pricing and costing using a fourth-generation language.

5. Critically analyze the benefits of automation for the company, both from a functional viewpoint (issues like flaw allowance, color matching, remnants, etc.), and from a practical viewpoint (no systems expertise in-house, adequacy of manual system, cost versus benefit).

6. Develop several automation strategies, and perform a cost/benefit analysis for each. Substantiate your analysis by visiting computer stores, hardware vendors or software consultants to obtain actual cost estimates.

7. Exhibits B1, B2, C1 and C2 contain costing/pricing information for a particular point in time. For example, exhibit B1 shows the purchase price of Allegro to be 15.04 pounds sterling, and B2 shows a standard landed cost of $34.50, and selling prices of $69.00 and $78.00. Normally companies want to maintain a history of such information. Could such a history be automated easily? If not, what are the alternatives? Discuss the limitations and benefits of automating historical data in terms of cost, amount of paper, ease of use and changes in methods of doing business.

North England Mills
Manchester, England

INVOICE: 60222

DATE: June 22

SOLD TO:

Quality Textiles International

123 Silver Creek Parkway

DESCRIPTION	QUANTITY ORDERED	OUR PATTERN	OUR COLOUR	UNIT PRICE	YARDS	TOTAL
B46M GEOMETRIC	1 pc	**Noranda** 21770	**Mauve** 289	L11.48	**52.5** 53	L 608.44

pkg
#817

TOTAL L 608.44

Exhibit A1: Invoice

North England Mills
Manchester, England

INVOICE: 60937

DATE: June 30

SOLD TO:

Quality Textiles International

123 Silver Creek Parkway

DESCRIPTION	QUANTITY ORDERED	OUR PATTERN	OUR COLOUR	UNIT PRICE	YARDS	TOTAL
E88B TAPESTRY	1 pc	**Allegro** 43228	**Aurora** 5	L15.04	**53.9** 55	L 827.70

pkg #823

TOTAL L 827.70

Exhibit A2: Invoice

North England Mills
Manchester, England

INVOICE: 61378

DATE: July 18

SOLD TO:

Quality Textiles International

123 Silver Creek Parkway

DESCRIPTION	QUANTITY ORDERED	OUR PATTERN	OUR COLOUR	UNIT PRICE	YARDS	TOTAL
M48A TWILL	1 pc	**Tyler** 21586	**Currant** 439	L17.50	**52.4** 52.	L 916.
B88D DIAG. STRIPE	1 pc	**Algoma** 21715	**Peach** 399	L11.48	**51.4** 52	L 596.96

pkg
#850

TOTAL L 1,512.96

Exhibit A3: Invoice

CUSTOMS IMPORTATION FORM

VENDOR: North England Mills
Manchester, England

ENTRY: Q326A153

SHIPPER: Conway Air Freight
New York, New York
10022

DATE: July 28

PKG	CONTENTS	FOREIGN COST	CONVERTED VALUE	DUTY RATE	DUTY
817	100% wool	608.44 UK	$1,010.01	25.0%	$252.50
823	100% wool	827.70 UK	$1,373.98	25.0%	$343.50
850	100% woven cotton	916.00 UK	$1,520.56	17.9%	$272.18
850	100% manmade fibers	596.96 UK	$ 990.95	25.0%	$247.38
	TOTAL VALUE		$4,895.50		
	TOTAL DUTY				$1,115.56

Exhibit A4: Customs importation form

PATTERN: 43228 ALLEGRO 54" wide @ L15.04

ORDER FROM: North England Mills
Manchester, England

OUR COLORS:	MILL'S COLORS:
Dawn	#7
Sunset	#3
Aurora	#5
Forest	#4
Painted Desert	#8
Autumn Haze	#15
Umber	#9

TERMS: net 30

SHIPPERS: Conway Air Freight,
APA Transport

Exhibit B1: Costing book - page 1 for Allegro

TYPE OF FABRIC: Flamestitch Tapestry
YARN CONTENT: 100% wool
DUTY RATE: 25%
PATTERN REPEAT: 14 1/8"

DATE	COST	RATE	OUR COST	DUTY	FRT	LANDS
Aug/22	15.04	1.75	26.32	6.58	1.60	34.50

PRICING:			
	B	69.00	100
	D	78.00	125

Exhibit B2: Costing book - page 2 for Allegro

PATTERN: 6415 FRESCO 54" wide @ 96 FF

ORDER FROM: Jean Dufresne
Paris, France

OUR COLORS:	MILL'S COLORS:
Frost Green	#101-12
Natural	#101-1
Navy	#108-14
Persimmon	#106-12
Sky	#109-1
Toast	#103-6
Vermouth	#102-7

TERMS: net 30

SHIPPERS: Pilot Freight,
Overseas Express

Exhibit C1: Costing book - page 1 for Fresco

TYPE OF FABRIC: Velvet Stripe
YARN CONTENT: 100% cotton
DUTY RATE: 15%
PATTERN REPEAT: none

DATE	COST	RATE	OUR COST	DUTY	FRT	LANDS
Aug/15	96.	.17	16.32	2.45	2.23	21.00

PRICING:			
	B	43.00	105
	D	48.00	127

Exhibit C2: Costing book - page 2 for Fresco

CODE	DATE	SUPPLIER	PATTERN	COST PER Y	EXCH RATE	OUR COST	DUTY RATE
313	June 30	Jean Dufresne	Fresco	96	.174	16.70	15%
314	July 1	North England	Allegro	15.04	1.80	27.07	25%
			Noranda	11.48	1.80	20.66	25%
			Tyler	17.50	1.80	31.50	17.9%
			Algoma	11.48	1.80	20.66	25%
⋮	⋮	⋮	⋮	⋮	⋮	⋮	⋮

Exhibit D1: Shipment log book - left side of page

DUTY COST PER Y	TOTAL YD's REC'D	TOTAL FRT COST	FRT COST PER Y	LANDED COST/Y NO FRT	ACTUAL LANDED COST		BL BOOK LANDED	BL BOOK FRT
2.51	89.6	428.23	5.38	19.21	24.59	X	21.00	2.23
6.37	53.9	437.20	2.07	33.44	35.51	X	34.50	1.60
4.81	52.5			25.47	27.54	X	26.90	1.60
5.19	52.4			36.69	38.76	X	36.25	1.60
4.81	51.4			25.47	27.54	X	26.90	1.60
	51.4							
	210.2							
⋮	⋮	⋮	⋮	⋮	⋮	X X X	⋮	⋮

Exhibit D1: Shipment log book - right side of page

PATTERN: Allegro **COLOR:** Aurora

PIECE NO: 739 **CODE:** 314

DATE RECEIVED: July 1 **COST/YD:** $35.51

FLAW ALLOWANCE BAL.					
FLAW USAGE					

DATE	CUSTOMER	YARDS USED	BALANCE REMAINING
July 1	received	--	53.9
July 2	Office Supply	30.9	23.0
July 5	Schneider Furniture	16.3	6.7

Exhibit E: Inventory card

Sold to: Office Supply

Date: July 2

Total Yards		Yards/Piece	Allowance
45.0	Allegro Aurora		
	PIECE: 616	14.1	—
	PIECE: 739	30.9	—

Exhibit F: Cutting order

PATTERN:	Allegro			
COLOR	**PIECE**	**YARDAGE**	**COST**	**VALUE**
Aurora	268	1.6	34.21	54.74
	317	3.2	34.90	111.68
	616	14.1	35.26	497.17
		18.9		663.59
Autumn Haze	67	8.2	33.16	271.91
Dawn	142	50.1	34.20	1713.42
	279	45.0	34.78	1565.10
		95.1		3278.52
Painted Desert	182	2.6	33.16	86.22
	53	1.2	31.20	37.44
	526	23.0	34.86	801.78
		26.8		205.44
Sunset	203	10.1	31.75	320.68
TOTALS		159.1		4740.14

Exhibit G1: Detailed inventory sheet

PATTERN	YARDS	VALUE	MARKED DOWN VALUE	AVG. VALUE PER YARD
Albi	101.7	743.43		7.31
Allegro	159.1	4740.14		29.79
Andros	445.9	5795.10		12.99
Ascot	1046.2	17,238.34		16.47
Bacon	321.4	5141.76		15.99
Berkley	247.5	3717.69		15.02
Brittany	375.5	3976.01		10.58
Butterfly	399.8	7525.45		18.82
Caprice	14.3	292.29		20.43
Chadwick	424.6	5616.03		13.22
Cord	809.9	10,412.98		12.85
Countess	506.4	7224.24		14.26
Crete	284.2	3581.33		12.60
Dauphine	469.7	4963.83		10.56
Domino	1504.5	32,346.79		21.50
Donegal	10.3	252.36		24.50
Dublin	22.9	499.96		21.83
Dunhill	46.2	774.77		16.76
Fina	430.9	6037.71		14.01
Floret	415.8	3392.10		8.15
Galway	390.0	8810.10		22.59
Glendowan	10.5	257.06		24.48
Granville	768.7	19,228.24		24.98

Exhibit G2: Summary inventory sheet

B.2 Custom Aircraft Corporation

Custom Aircraft Corporation is in the business of manufacturing custom-made aircraft. They also distribute aircraft-related products for a number of other manufacturers. The company's current sales catalog contains 458 items, of which approximately half are manufactured by the company. The company's customers include wealthy individuals and corporations who wish to purchase custom-made executive jets. The company has a commissioned sales force who receive commissions at the rate of 10% on items manufactured by the company (manufactured items), and 8% on items distributed by the company (distributed items). Each sales representative has an exclusive territory and customer list.

At the present time, the company's procedures are entirely manual. However, management have decided to create a data processing function, beginning with the products division. They have hired a systems analyst and formed a data processing department for the purpose of developing and implementing an automated system.

The new systems analyst prepared the following operational profile and summary from interviews conducted with several levels of users at the company. The analyst began the summary by tracing a customer order through the six departments of the product division.

B.2.1 Order Processing

Each order must be accompanied by a 25 percent deposit, and commercial customers must have a favorable Dun and Bradstreet report. All orders are directed to the order processing clerk who carries out the following procedures:

- The clerk determines if the order is from an old or new customer by referring to the customer file.
- If the order originates with an old customer, then the customer's name and address are verified and the file is updated with any changes.
- The three credit references required of each customer are contacted to determine if the customer's financial position has changed.
- If the customer is new, the file is updated with the customer's name, address, three credit references and phone number.

- Each item number, description and price is verified with the product catalog.

- A sales commission notice is written up and sent to payroll. In the notice, manufactued items and distributed items are grouped and totalled separately to reflect the different rate of commission payable on each group.

- The verified order is sent to inventory control.

B.2.2 Inventory Control

Inventory control is responsible for maintaining reasonable levels of the company's inventory. Inventory includes all raw materials, purchased items, components and finished products. Every item, regardless of type (purchased or manufactured) has a maximum and minimum quantity assigned to it. Management adjust these maximum and minimum levels from time to time, as required. Each time the inventory control clerk updates an item in the inventory file after a shipment has been delivered to a customer, the remaining quantity on hand must be checked to see if it is less than the minimum quantity. If it is, the following actions are taken:

- If the item is a manufactured item, then the inventory control clerk issues a produce inventory notice and sends it to the production scheduling department.

- Otherwise (for purchased goods) the clerk sends a purchase inventory notice to the purchasing department.

When a shipment of raw materials or purchased items is received, each pertinent item in the inventory file is updated. If the quantity on hand exceeds the maximum quantity, the clerk takes the following action:

- If the item is a manufactured item, it is necessary to issue an overstock notice to both production scheduling and to management.

- For purchased items, an overstock notice is issued to management.

When a new order reaches inventory control, the following actions take place:

- The order is rewritten into sub-orders, one for each item.

- The order is filed in the orders pending file.
- Each sub-order is checked against the inventory file to determine if the sub-order can be filled from items presently in stock.
- If the sub-order can be filled from current inventory levels, then
  - the inventory is adjusted by the required amount
  - a shipping slip is prepared and sent to the warehouse
  - the sub-order is filed along with a copy of the shipping slip and the original order in the orders pending file.
- If the sub-order cannot be filled from current inventory, then
  - if the item in the sub-order is a manufactured item, the sub-order is filed in the manufacturing suspense file and a produce order notice is written and sent to production scheduling.
  - If the item is a distributed item, the order is filed in the distribution suspense file and a purchase order notice is sent to purchasing.

Production scheduling attaches the produce order notice to a production schedule and forwards it to manufacturing. When the item has been completed, the notice is marked up to indicate the date of completion, and both the item and the notice are sent to the warehouse. The receiving clerk in the warehouse allots a storage location to the item and indicates this location on the notice. Then the notice is sent back to inventory control, where the inventory control clerk

- enters the information into the inventory file, then adjusts the records accordingly
- activates the sub-order by retrieving it from the manufacturing suspense file and attaching the marked-up order notice to it
- writes up a shipping slip and sends it to the warehouse
- files the suborder, notice, order and a copy of the shipping slip in the orders pending file.

The inventory control clerk issues a produce inventory notice when the on-hand level of any manufactured item falls below the minimum quantity. The procedure for this is identical to that for a produce order notice described above. The reason for having two different notices is as follows: production scheduling must give scheduling priority to any manufacturing required to fill a customer's order over manufacturing which is needed to fill inventory reserves. The two different types of notices allow production scheduling to distinguish between the two requirements.

When inventory control once more receives the produce inventory notice, it indicates that the requested items have been produced and located in the warehouse. The inventory file is then updated accordingly.

A receiving slip is prepared by a receiving clerk in the warehouse to indicate the receipt of items from suppliers. Items received could be either distributed items, or raw materials and purchased parts for manufacturing.

When Inventory control receives a receiving slip, the following action is taken:

- The received items are checked against the distribution suspense file to determine if any suspended sub-orders can be activated.
- If a sub-order can be filled, it is activated by removing it from the distribution suspense file and attaching the receiving slip.
    - both the incoming shipment and adjustment amount are noted
    - a shipping slip is prepared and sent to the warehouse
    - all the documents (sub-order, receiving slip, order) are attached to the original verified order and filed in the orders pending file.
- If a sub-order cannot be filled, the inventory file is updated with the appropriate amount.

Before production scheduling can send the daily schedule to manufacturing, it must determine if the raw materials or components are available. To this end, a materials and components list is prepared and sent to inventory control where it is checked against current inventory levels.

If any item on the list is out of stock, the entire list is returned to production scheduling, and the missing item(s) ordered. Inventory control does not update inventory records or take steps to have the raw materials and components delivered to manufacturing until it receives a materials and components list which it can completely fill.

The steps taken by inventory control upon receiving a materials and components list are as follows:

- If all items are in stock, then
  - production scheduling is notified so that the schedule can be sent to manufacturing
  - the inventory records are updated according to the items on the list
  - a picking slip is written up and sent to the warehouse where the required items will be "picked" out of inventory and delivered to manufacturing.
- If any item on the list is not in stock, the whole list is returned to production scheduling so that a new schedule can be prepared.
- If the out-of-stock item is a distributed item, raw material or purchased part, a purchase order notice is written up and sent to purchasing.
- If the out-of-stock item is a component, then a produce order notice is sent to production scheduling.

B.2.3 Production Scheduling

Production scheduling is responsible for keeping manufacturing working to capacity at all times. It determines the schedule according to a set of priorities which have been established by management. A major function of this department is to create the production schedule for manufacturing, a task which is performed each day. It works one day in advance of manufacturing, so on Monday, for example, it creates Tuesday's schedule. The priorities are as follows:

- First priority is given to the manufacture of items on produce order notices because these are items that have been ordered by customers.

- Second priority is given to items on produce inventory notices. These are items that have fallen below minimum inventory levels.

- Any remaining manufacturing capacity is applied to building up inventories of manufactured items and components according to a list furnished by management which is based on best selling manufactured items.

Before a production schedule is sent to manufacturing, production scheduling determines, through the interface with inventory control, described above, whether sufficient raw materials and components are available.

At about 1:00 P.M. each day production scheduling assembles all notices received during the morning and previous afternoon, and prepares a production schedule according to the priorities set out above. Generally the procedure has the following pattern:

- The production schedule is prepared.

- Next the materials and components list is created after querying the manufacturing specification file.

- The materials and components list is sent to inventory control where availability of raw materials and components is checked.

- If any item on the list is out of stock, then the first three steps above are repeated after the schedule has been amended to avoid the effect of out-of-stock materials or components.

- If all items on the materials and components list are in stock, the produce order and produce inventory notices are attached to the production schedule and it is sent to manufacturing.

- If the production schedule permits the manufacture of extra manufactured items and components (as described above), then production scheduling prepares additional produce inventory notices for these items.

B.2.4 Manufacturing

Production begins each morning at 7 A.M. according to the production schedule received the previous afternoon. The raw materials and components required are delivered the previous afternoon by the warehouse workers.

The appropriate notice (for producing orders or inventory) is attached to each item produced. The completion date is indicated on the notice, and the completed item and the attached notice are sent to the warehouse.

B.2.5 Warehousing

Warehousing, in addition to controlling the physical storage space of the company, handles all shipping and receiving of items.

Shipping retrieves items from storage and delivers them to customers. (Pilots are hired on a contract basis as the need arises.) Receiving accepts deliveries from outside suppliers, and places these items into storage. In addition, both shipping and receiving interact with manufacturing. Shipping delivers raw materials and components to manufacturing each day, while receiving accepts completed manufactured items and components from manufacturing and places these items in storage.

When the warehouse (shipping) receives a shipping slip from inventory control, a clerk retrieves the items from storage, packages them, and ships them to the customer via common carrier. (For delivery of planes, the services of a local pilot are used.) When the items leave the shipping dock, the time and date are noted on the shipping slip. One copy is filed in the shipping file. A second copy of the slip is sent to accounts receivable, so that an invoice can be prepared.

The shipping clerk retrieves raw materials and components, as indicated on the picking slip, and delivers these items to manufacturing ready for the following day. The time and date are indicated on the slip, and it is filed in the shipping file.

The purchasing department sends a copy to receiving of every purchase order issued to outside suppliers. These copies are filed in receiving's purchase order file.

When items arrive that have been ordered from suppliers, they are checked against the corresponding purchase order. Next, the items are placed in a storage location in the warehouse, and a receiving slip is prepared, with the location noted. The slip also includes the item number and quantity of all receipts. One copy of the slip is sent to accounts payable, where it is matched with the supplier's invoice when it arrives. The other copy of the receiving slip is sent to inventory control, where the inventory file is updated. Sub-orders held in the distribution suspense file may be activated as a result of the updating.

After the manufacture of various items has been completed, such items are sent to the warehouse, along with either a produce order notice, or a produce item notice. The location of each item is indicated on the notice, and then the notice is forwarded to inventory control, and the inventory file is updated.

B.2.6 Purchasing

The purchasing department responds to purchase order notices and purchase inventory notices. These two types of notices distinguish items being purchased to fill a customer's order, and items being purchased to build up inventory levels. This department has standing instructions to give priority to those purchases made to fill customers' orders.

Management negotiate contracts with all suppliers, for each and every item in inventory that is not manufactured by the company. These contracts are the responsibility of management. The contracts are filed in the supplier's contract file, and can be retrieved when required.

The following actions are taken when a purchase order notice or purchase inventory notice is received:

- The supplier's name, address and contract are retrieved from the supplier's contract file.
- A purchase order is prepared, and sent to the supplier.
- Additional copies are sent to accounts payable, receiving, and a final copy is filed in the purchase order file.

B.2.7 Possible Assignments

1. Analyze the system described above, and develop a conceptual data model, using an entity-relationship diagram.
2. Write down examples of business functions from each of the six areas described above (order processing, inventory control, production scheduling, manufacturing, warehousing, purchasing) that your E-R model could support. One example might be accept/reject/modify production schedule.

3. Draw a complete set of physical and logical data flow diagrams for Custom Aircraft. The physical diagrams should have at least three levels, and the logical diagrams at least five levels. Supply any missing elements, and note your assumptions. (Suitable for groups of two to three.)

4. Reconcile the conceptual data model and the DFD, modifying them where necessary.

5. Develop a logical data model, using your (modified) E-R diagrams as the basis.

B.3 Omnitech Corporation

The Omnitech Corporation manufactures electronic components for the consumer market. Its sales, currently $75 million per year, are projected to be approximately $100 million within the next two years. Its product line consists of two hundred products, including variations for customers whose requirements differ because of the markets they serve. Certain products are shipped directly from the manufacturing plants to large-volume customers; however, customer orders are usually filled from one of the five field warehouses, strategically placed throughout the country.

Corporate headquarters are centrally located in St. Louis; the locations of the company's manufacturing plants are Minneapolis, Philadelphia and Los Angeles. Each manufacturing plant has an attached warehouse where finished goods are held temporarily for shipment to customers, or for shipment to field warehouses. Currently the firm employs approximately 2,500 workers.

B.3.1 Marketing

The main functions of marketing are to contact potential customers and to sell the company's products through salespeople, distributors, advertising and special promotions. If products are sold by salespeople, a multicopy sales order is originated by either the individual salesperson or the order group of marketing. Copies of a sales order are distributed as follows: the customer gets the original, and duplicates go to the salesperson, manufacturing, stock control, shipping, and accounting. Marketing may also prepare other forms, such as contracts, bids, back orders and change orders.

B.3.2 Research and Development, Engineering

Often marketing receives inquiries regarding a new product, and will initiate a research and development (R & D) feasibility study, provided that market prospects look promising. As well, if marketing research develops an idea for a potential product, an R & D feasibility study will be conducted. In the past, R & D has played a crucial role in the company's growth rate, and it is expected to continue doing so.

Before the manufacturing process can commence, product prototypes must be developed and tested, and final product design must be specified and accepted by marketing. Engineering specifications are developed and form the basis for producing the manufacturing requirements for such products: raw materials and purchased parts, components, finished goods. The requirements for the product to be manufactured are summarized on a *bill of materials*. A bill of materials is used to determine the number of detailed items needed in the production process. With the engineering specifications and appropriate bills of materials prepared, production can be scheduled.

B.3.3 Manufacturing

The manufacture of products involves many steps. Not only must plant, equipment and work stations be provided, but also appropriate personnel hired and trained to utilize the manufacturing facilities. Raw materials must be planned for and purchased to arrive just in time for use in production. Production must be scheduled, work in progress tracked and monitored, and completion dates met. As well, the accurate costing of goods (raw materials, labor, workstation) is crucial to profitability. Effective management in this area is vital.

The company's products are produced in anticipation of demand, in response to firm customer orders, or a combination of both. If the goods are being produced to order, the sales order becomes the basis of the production schedule. The usual arrangement is to have the production planning group responsible for this.

Manufacturing materials requirements can be obtained from inside or outside the company, and for materials required from suppliers, purchase orders are written. If materials are available from within the company, the warehouse forwards the required items to manufacturing.

B.3.4 Purchasing and Inventory

Purchasing is concerned with procuring raw materials and purchased parts, equipment, and supplies, as well as other products and services required to ensure continued operation. Contracts are required for all desired goods and services. If a service or product is desired for which there is no contract, requests for quotation are sent to several prospective suppliers, and from the responses, a supplier is chosen.

When goods are received from suppliers, and verified by receiving, they are approved for payment. Goods, including inventory and supplies for internal use, are delivered to the appropriate destination within the company. Stock control must replace stock when minimum levels are reached.

B.3.5 Distribution

After a customer's order has been manufactured, or the goods are available for shipment to the field warehouse, they are carefully packed, labeled, and transported. Shipping orders authorize delivery. When shipments are made via common carriers, a bill of lading, a contract between the consignor and the carrier, is required.

B.3.6 Accounting and Finance

After the shipment of finished goods, the customer account is prepared. Normally, because of the large costs of production, customers are required to pay 25 percent of their orders in advance. Depending upon the terms of payment, the balance may be received in 10, 30 or 60 days after delivery. Payments are deposited in the company's bank in St. Louis.

In addition to billing and collecting, accounting and finance are concerned with disbursing funds and projecting cash requirements. Disbursement of funds occurs for things such as payroll, purchase of goods and services, and for the company's line of credit. Labor and overhead distribution charges are made to various departments.

The government requires a variety of financial information from the company. Each month, the company forwards the income tax amounts which have been deducted from employees' earnings. In addition, there are reports on social security deductions, federal and state unemployment compensation reports and state income tax amounts. Other governmental information on which statistical data for businesses in the U.S. is based, is required of the company.

B.3.7 Possible Assignments

1. Using the preceding description of the Omnitech Corporation, develop a conceptual data model of the company, and diagram using an E-R diagram.

2. Diagram the functions of Omnitech in a leveled set of logical data flow diagrams. Validate your DFD's against your conceptual model.

3. Using your results above, develop a relational database for Omnitech.

B.4 Federal Supply Company

The following observations were made by the systems analyst as part of a study of the Federal Supply Company.

Federal Supply Company manufactures custom-made metal and plastic parts and components. All parts and components are made to order on receipt of a purchase order from customers. When the required parts have been completed, they are immediately shipped to the customer. Approximately 75,000 customer-initiated orders are processed by the company during a single year. A typical order contains, on the average, a request for five different items, in quantities of 12 to 2000 each.

B.4.1 Order Processing

When customer orders are received in the order processing office, an entry is made in a transaction log book, noting the customer name, date of receipt, and order number, if any. The purchase orders are then sent to the accounting office for credit verification. This takes about one day. If credit is approved, the order is returned to the order processing office where the log is updated and a work order is prepared. If credit is not approved, the order is returned to the customer by the accounting office along with a letter indicating the reason for denying credit. A pink sheet (credit action form) is sent to the order processing department indicating the credit decision. The pink sheet is used to update the transaction log book and is then discarded.

The work order is prepared in three copies. One copy is attached to the customer's order and is placed in the active sales file. A second copy is sent to inventory control, while the third copy is forwarded to the manufacturing office, to be used in scheduling production.

B.4.2 Inventory Control

The inventory control copy of the work order is used to prepare a materials requisition on the basis of information contained in the bills-of-material file. This is a file that contains a breakdown of all parts and materials used in the manufacture of a specific item. The resulting purchase requisition, which itself is prepared in three copies, is sent to the purchasing department, with one copy remaining in the inventory control department in the active purchase requisition file. Another copy is forwarded to manufacturing.

B.4.3 Purchasing and Receiving

When the purchasing department receives the requisition, it prepares a formal purchase order and sends three copies to the vendor. Fourth, fifth and sixth copies are placed in the active purchase order file, in the manufacturing office's active file, and in an accounts payable file, respectively.

When materials are received in the shipping room, the incoming packing list (two copies) is removed from the shipping carton, checked for accuracy (did all of the ordered items arrive?), and copies are sent to purchasing and manufacturing, respectively. When purchasing receives its copy of the packing list, it is compared to the original purchase order. If all ordered merchandise has been received, the purchase order copy is placed in the closed purchase order file and the packing list is sent on to the accounting department.

Upon receipt of an invoice from the supplier, the amount is paid by check. When the bill has been paid, the packing list, accounting's copy of the purchase order, the invoice and a "paid by check" form are placed in the closed accounts payable file.

B.4.4 Manufacturing

The copy of the packing list manufacturing receives is paired with the purchasing requisition and purchase order, and production is scheduled. After the merchandise has been manufactured, it is sent to the shipping area together with the purchase requisition. The shipping office prepares a packing list (in triplicate) and ships the merchandise to the customer along with two copies of the packing list.

B.4.5 Accounts Receivable

The third copy of the packing list is sent to the accounts receivable department where an invoice is prepared and sent to the customer. A copy of the invoice is stapled to the packing list and placed in the active accounts receivable file. When payment has been received from the customer, the documents are moved from the active to the closed accounts receivable file.

B.4.6 Possible Assignments

1. Using the preceding description of Federal Supply, develop a conceptual data model of the company, and document it using an E-R diagram.

2. Diagram the functions of Federal Supply in a leveled set of logical data flow diagrams. Validate your DFD's against your conceptual model.

3. Using your results above, develop a relational database for the company.

B.5 Sweet Shops Incorporated - Catalog Division

The Sweet Shops Catalog Division is a division of Sweet Stores Inc., and is concerned with the mail-order catalog operation. The catalogs are widely distributed to the general public, and are published each spring and fall. They contain thousands of items of the type found in large department stores.

B.5.1 Orders

Customers make purchases by mailing their orders, either with a prepayment or with a Sweet's credit card number to the centralized mail-order processing department, located in Elmira, Ontario. The orders are filled, if possible, from one of Sweet's warehouses located near all major population centers across the country.

Orders are checked against inventory to determine if the ordered items are in stock. If they are, the quantity on hand is adjusted by the number of ordered items, and a picking slip is sent to the appropriate warehouse. If the ordered items are not in stock, an out-of-stock notice is sent to the customer. The company will ship any partial orders it can, and does not hold backorders.

B.5.2 Prepayment and Credit

Two types of orders are processed by the company: prepaid orders, where the order is accompanied by cash, check or money order, and credit card orders.

There are two types of prepaid orders: those which include prepayment initially, and those which send prepayment after having been denied further credit on their credit cards. When prepayments are received, they are recorded in the accounts receivable file and are deposited in the company's bank account.

The company policy regarding credit is that they will extend credit only to holders of Sweet's credit cards, and only if the balance on their account plus the amount of the new order does not exceed a preassigned credit limit. If the credit is not approved, the order is put into suspense and a request for prepayment is sent to the customer.

B.5.3 Purchasing

Purchasing is concerned with the procurement of inventory and supplies required to meet company needs. Procurement commences with the completion of a purchase requisition, prepared in duplicate. One copy is forwarded to purchasing, one is retained by the originating department.

A purchase order is prepared and mailed to the vendor. Purchasing keeps one copy. It contains the items to be shipped, prices, terms of payment and shipping conditions. The original copy of the purchase order is sent to the vendor and duplicate copies are distributed to receiving, accounts payable and the originating department.

B.5.4 Receiving

When goods are received from suppliers, the receiving clerk checks them against the purchase order and prepares a receiving report noting any discrepancies between the order and what was received. Copies of the receiving report are sent to purchasing and accounts payable. Receiving keeps one copy.

B.5.5 Inventory Control

Inventory records are maintained by inventory control who are responsible for replacing inventory when it reaches the reorder level. They prepare a purchase requisition to reorder goods.

B.5.6 Accounting and Payroll

Accounting prepares customer invoices after goods have been mailed to customers. Two copies of the invoice are sent to the customer, another is filed in the invoice file and the fourth copy is placed in the customer's file folder. The amount of the invoice is noted on the customer's account and payments are recorded in the cash receipts journal.

Accounting pays suppliers' invoices after the invoice is compared with the purchase order and receiving report. They prepare a check and voucher. The voucher notes details such as purchase order number, supplier number and discount.

Payroll, a section of the accounting department, is responsible for preparing employees' pay checks and maintaining payroll records. Timecards are used by all employees. A weekly timesheet is prepared from the timecards, and from it, individual payslips are prepared. Gross pay is calculated by referencing the rate-of-pay file. Income tax, unemployment insurance deductions, social security deductions and net pay are noted on the payslips. The company pays workmen's compensation premiums.

Employees are paid weekly. Each month checks are sent to the following agencies:

- The Internal Revenue Service, for income tax deductions
- Unemployment Insurance Commission, for company and employee contributions
- Social Security Commission, for company and employee contributions
- Worker's Compensation Board, for company contributions

The individual pay checks are made up from payslips, and employees receive both in an envelope.

B.5.7 Possible Assignments

1. Using the preceding description, develop a conceptual data model of the described portion of the company, and document it using an E-R diagram.
2. Diagram the functions of the Catalog Division in a leveled set of logical data flow diagrams. Validate your DFD's against your conceptual model.
3. Using your results above, develop a relational database and document it using data structure diagrams.

INDEX